T0291034

Rainer Dahlmann (Ed.)
Failure Analysis for Plastics Products

Also of interest

Plastics in the Circular Economy
Vincent Voet, Jan Jager and Rudy Folkersma, 2021
ISBN 978-3-11-066675-5, e-ISBN (PDF) 978-3-11-066676-2

Infrared Spectra of Rubbers, Plastics and Thermoplastic Elastomers
Yvonne Davies, Jason Davies and Martin J. Forrest, 2019
ISBN 978-3-11-064408-1, e-ISBN (PDF) 978-3-11-064575-0

Biopolymers.
Environmental Applications
Edited by: Jeyaseelan Aravind and Murugesan Kamaraj, 2023
ISBN 978-3-11-099872-6, e-ISBN (PDF) 978-3-11-098718-8

Fiber Materials.
Design, Fabrication and Applications
Edited by: Jeenat Aslam and Chandrabhan Verma, 2023
ISBN 978-3-11-099274-8, e-ISBN (PDF) 978-3-11-099289-2

Polymer Engineering
Editet by: Bartosz Tylkowski, Karolina Wieszczycka, Renata Jastrząb
and Xavier Montane, 2022
ISBN 978-3-11-073844-5, e-ISBN (PDF) 978-3-11-073382-2

Failure Analysis for Plastics Products

Methods and Instruments for Troubleshooting and
Remedial Measures

Edited by
Rainer Dahlmann

DE GRUYTER

Editor
Prof. Dr. rer. nat. Rainer Dahlmann
Institute for Plastics Processing
at RWTH Aachen University (IKV)
Seffenter Weg 201
52074 Aachen
Germany
rainer.dahlmann@ikv.rwth-aachen.de

ISBN 978-3-11-078562-3
e-ISBN (PDF) 978-3-11-078564-7
e-ISBN (EPUB) 978-3-11-078569-2

Library of Congress Control Number: 2024935549

Bibliographic information published by the Deutsche Nationalbibliothek
The Deutsche Nationalbibliothek lists this publication in the Deutsche Nationalbibliografie; detailed
bibliographic data are available on the internet at http://dnb.dnb.de.

© 2024 Walter de Gruyter GmbH, Berlin/Boston
Cover image: Institut für Kunststoffverarbeitung in Industrie und Handwerk an der RWTH Aachen
Typesetting: Integra Software Services Pvt. Ltd.

www.degruyter.com

Preface

Damage to plastic products is a daily occurrence. The reason for this is not a weakness of the material class of plastics itself, but rather the enormous variety of stresses resulting from the fact that plastics can be found in all areas of application. Sometimes the causes and conditions under which damage has occurred are obvious, which means that avoiding this damage in the future is only a small step. Experience is a great help here. Very often, however, the framework conditions are complex, the number of interests involved is very multi-layered and so opaque that the reconstruction of a damage process – and thus the identification of a cause – is not possible without further ado. And in most cases, damage cannot be attributed to a single cause, but is multi-causal; there is a number of influences that favour damage without being the cause in isolation.

If the investigation of damage is carried out as a service, a methodology and suitable tools are required, as the damaged products, their properties, and detailed functionalities are generally not well known to the damage analyst. They often first familiarise themselves with a product on the basis of a damage image and then have a strong need for information about its development and history before they can focus on the damage process itself. Based on such situations and the necessary structuring, methods, and tools for systematic damage analysis have been developed over decades at the Centre for Plastics Analysis and Testing at the Institute of Plastics Processing in Aachen, Germany. This development has always been harmonised with experts from other institutions who regularly exchange information in the expert group of the Association of German Engineers (VDI), "Failure Analysis and Prevention for Plastic Products".

This book endeavours to share the essence of decades of practical experience and to provide ways to accelerate learning from damage to plastic products. It is aimed at product, production, and quality managers as well as experts and assessors with expertise in the field of plastics technology.

I would like to thank all the authors for their willingness to share their experiences and to write them down in such a concise form. I would also like to express my sincere thanks to the other employees of the Centre for Plastics Analysis and Testing, who do not appear here as authors, but historically have also played a decisive role in the development of the contents. My thanks also got to the employees of the publishing house for their constructive cooperation and Edge Fischer, who, in addition to her content contribution, has thankfully organised and monitored the organisational and editorial processes among the authors and managed the communication with the publishing house.

https://doi.org/10.1515/9783110785647-202

All the work and results contained in this book were carried out at the Institute of Plastics Processing, Aachen, Germany. I am grateful for the freedom to organise this work, especially to Christian Hopmann, who helped to set the appropriate framework already many years ago.

Rainer Dahlmann

Contents

List of authors

Jan Buir
Institute of Plastics Processing IKV
Seffenter Weg 201, 52074 Aachen, Germany
Jan.buir@ikv.rwth-aachen.de
Chapters 4, 5

Sina Butting
Institute of Plastics Processing IKV
Seffenter Weg 201, 52074 Aachen, Germany
Sina.butting@ikv.rwth-aachen.de
Chapter 5

Hakan Çelik
Institute of Plastics Processing IKV
Seffenter Weg 201, 52074 Aachen, Germany
Hakan.celik@ikv.rwth-aachen.de
Chapter 4

Tobias Conen
Institute of Plastics Processing IKV
Seffenter Weg 201, 52074 Aachen, Germany
Tobias.conen@ikv.rwth-aachen.de
Chapters 3, 5

Prof. Dr. rer nat. Rainer Dahlmann
Institute of Plastics Processing IKV
Seffenter Weg 201, 52074 Aachen, Germany
Rainer.Dahlmann@ikv.rwth-aachen.de
Chapters 1, 2, 3, 4, 5

Edge Fischer
Institute of Plastics Processing IKV
Seffenter Weg 201, 52074 Aachen, Germany
Edge.fischer@ikv.rwth-aachen.de
Chapters 4, 5

Matthias Klimas
Institute of Plastics Processing IKV
Seffenter Weg 201, 52074 Aachen, Germany
Matthias Klimas@ikv.rwth-aachen.de
Chapter 5

Michèle Marson-Pahle
Institute of Plastics Processing IKV
Seffenter Weg 201, 52074 Aachen
m.marsonpahle@gmail.com
Chapters 4, 5

Meike Robisch
Institute of Plastics Processing IKV
Seffenter Weg 201, 52074 Aachen, Germany
Meike.robisch@ikv.rwth-aachen.de
Chapters 4, 5

Dr. rer. nat. Sabine Standfuß-Holthausen
Institute of Plastics Processing IKV
Seffenter Weg 201, 52074 Aachen, Germany
Sabine.Standfuss-Holthausen@ikv.rwth-aachen.de
Chapters 4, 5

Christiane Wintgens
Institute of Plastics Processing IKV
Seffenter Weg 201, 52074 Aachen, Germany
Christiane.wintgens@ikv.rwth-aachen.de
Chapters 4, 5

Christoph Zekorn
Institute of Plastics Processing IKV
Seffenter Weg 201, 52074 Aachen, Germany
Christoph.zekorn@ikv.rwth-aachen.de
Chapters 2, 4, 5

https://doi.org/10.1515/9783110785647-204

1 Introduction

Fault and damage analysis is an important tool for product safety, quality assurance and avoiding economic losses. The prerequisite for success is the reliable mastery of the necessary tools. This book is intended to provide a sound insight into the systematic approach to failure and damage analysis of plastic products and serve as a guide to practical implementation.

The procedure for determining the causes of damage events utilises scientific methods in many respects. This essentially concerns the use of instrumental analysis with all facets of preparing, carrying out and interpreting these investigations. The recording and description of a damage pattern and environment must also fulfil scientific requirements in that these are largely the results of subjective observations, which require a clear demarcation from the interpretation of the observed in the documentation of the damage. After all, embedding what has been observed in already existing generally recognised scientific findings and correlations as a basis for the formation of valid damage theses is a scientific creative process. In practice, however, the scientific claim of a damage analysis usually ends here because the refutation of the damage hypotheses formed, which in Popper's sense represents a gain in knowledge, is not consistently carried out. As a rule, for economic reasons, one has to be content with the damage theses that are still valid at the end of the damage analysis. In this respect, systematic damage analysis as a whole is not a science – it does, however, make use of its methods in order to be purposeful and successful.

This book is intended to provide a practical overview of the special procedure such as scrutinising the damage environment, the well-founded development of damage theories and the definition of remedial measures. The importance of the chronology of action and essential technical terms are named and explained in the book. Special attention is paid to the complex interplay of design, processing, material properties and material reactions to external influences. In particular, manufacturing and materials engineering aspects are also considered so that systematic failure analysis offers a comprehensive and universal method for determining the causes of damages of all types of products. In addition to material technology and analytical principles relating to the topic of failure and damage analysis on plastic products, various quality management tools and the systematic procedure for determining damage and the resulting damage prevention are covered in detail.

The book is divided into five chapters. Chapter 2 is dedicated to the fact that (particularly complex) damage requires good structuring in order to be analysed. The methodology presented there is closely related to the VDI 3822 standard and takes into account the specific characteristics of plastic products, which in most cases originate from automated series production. This aspect in particular is highlighted in Chapter 2, as it is often possible to narrow down the problem with just a few decisive questions. Chapter 3 deals with the basics of plastics technology. This chapter is not

https://doi.org/10.1515/9783110785647-001

intended to provide laypersons with sufficient knowledge of plastics technology to be able to carry out damage analyses independently, but rather to enable experts to familiarise themselves with the interrelationships presented and the background and terminology used in the book. Chapter 4 provides a brief summary of what we consider to be the most important instrumental analysis methods for analysing damage to plastic products. The descriptions are relatively brief, but there are references to where more detailed information can be found. Furthermore, typical application examples are given from the perspective of analysis methods. In particular, process and structural simulations as methods, which have gained in importance in failure analysis in recent years, are also addressed here. Finally, Chapter 5 provides examples from practice, which are presented here in a highly structured form to make it as easy as possible to browse through them. We have not set ourselves the task of describing the true complexity of the damage cases here, but merely to convey the most important framework conditions and findings on the basis of the structure.

Determining the causes of damage to plastic products is always a challenge due to the wide range of possible influences. However, damage also often represents a unique opportunity to gain valuable insights that are otherwise inaccessible and to incorporate these into research and development. Then you can learn from damage.

Rainer Dahlmann, Christoph Zekorn

2 The methodical approach in failure analysis

2.1 Introduction

Failures to plastic products usually occur largely unforeseen but require immediate problem-solving action. Depending on the extent of the damage and, above all, its foreseeable consequences, damages generate a great deal of attention and immediate pressure to act. As a rule, attention and pressure increase with the current development stage of a component or product along the value chain. Defects or unscheduled test failures can still be handled comparatively well during the evaluation of prototype or pre-series components and can be controlled with moderate effort. In real component field use, however, these can quickly become more widespread and cause no inconsiderable collateral damage. Damage events therefore quickly lead to forms of actionism among those directly responsible, and the search for the cause can be deliberately steered in certain and possibly wrong directions, motivated by the vested interests of the persons or parties involved. Systemic failure analysis provides tools to handle such situations appropriately while maintaining an objective view. It counters the above-mentioned blind actionism with a holistic approach and a thesis-based procedure and offers benefits that go beyond (Figure 2.1).

Figure 2.1: Benefit of failure analysis.

Conventionally, the term failure analysis is used in the context of damage caused by misuse of products. An example is the analysis of accidental damage to motor vehicles, which focuses strongly on the cause of the damage and the determination of the amount of damage. Systemic failure analysis goes well beyond these points of view. It considers all influences that can theoretically and from the view of the individual case lead to the present damage picture. This includes, among other things, all aspects of the design, including the definition of the product requirements, up to and

https://doi.org/10.1515/9783110785647-002

including the conceivable influences on the product during use. In particular, it considers all manufacturing and material aspects and thus becomes a comprehensive and universal method for finding the causes of products of any kind.

This chapter describes the procedure for failure analysis, especially for products made of plastics. On the one hand, it can be stated that the basic structure of the procedure primarily follows a logic that is completely independent of the material from which the defective product was made. However, at the point of a damage analysis where a product state must be described on the basis of internal properties, it becomes clear that material-specific methods must be used for this purpose. A typical example is structural analysis: this requires completely different instrumental analysis methods for metals and ceramics than for plastics.[1]

But there are further causes why the material specific distinction is necessary: There are fundamentally different manufacturing processes, and they exhibit completely different failure mechanisms. Furthermore, failure to plastic products often has a higher degree of complexity. This is due to the interaction of different circumstances:

- Plastics are always a composition of many components (polymers or polymer blends, additives, fillers, reinforcing materials), whose exact consistency is only known by the raw material manufacturer. The failure analyst does not have access to this information.
- Plastic materials are highly diversified. This makes it a challenge to evaluate the target condition of a plastic component on the basis of a failure pattern alone. In practice, therefore, reference products are always sought in the failure environment which is intended to represent the target condition in some way. As simple as this may sound at first, it turns out to be one of the major challenges in practice.
- Plastic products are very often mass-produced. Failure parts represent at least initially an unknown part of the total production, which in turn may be divided among different batches, production sites or suppliers. Thus, the entire product life cycle from the definition of product requirements to use must be considered first and foremost.
- The design of plastic products and their implementation in mould technology is subject to a wealth of competing and sometimes contradictory requirements. For example, design demands placed on a plastic moulded part often have to be reconsidered and adapted in terms of a design suitable for the plastic (e.g. radii and

---

1 The Association of German Engineers (VDI) has created a guideline VDI 3822 for Failure Analysis already many years ago [60]. It basically distinguishes between damage analyses on metallic, thermoplastic and elastomeric products, for each of which separate series of guidelines have been developed. The distinction takes account of the fact that these material groups are subject to fundamentally different manufacturing processes, exhibit different damage mechanisms and require different analytical methods for their detection.

wall thicknesses) or possible manufacturing processes (compatible mould complexity). Component and mould design therefore always represent a compromise, especially in the case of sophisticated products, which necessarily represents a non-optimal implementation for some requirements.

- The manufacture of plastic products is generally subject to high economic pressure. Here, too, the converter must find a compromise that ensures good product quality with satisfactory cycle times or process times, respectively. The fact that an outwardly or apparently good product quality is often used or must be used as a sufficient evaluation criterion on the production line, on the basis of which cycle time optimisation then takes place, creates the breeding ground for damage analyses that become necessary later.

The overall result is that damage to plastic products is rarely monocausal and quickly reaches a high degree of complexity. As a rule, there are several influences that promote damage, which have to be weighed up and balanced against each other on the basis of available information and empirical values. Systemic damage analyses must therefore follow a holistic approach, which allows for an in-depth examination of details as well as a return to a superordinate, objective level of observation. This minimises the risk of inadvertently disregarding information that is important for damage assessment. The methodology of systemic damage analysis can also be applied to much broader problems if the concept of damage is extended to include unexpected obstacles in development, pre-production or production that give rise to the expectation of damage. The methods of systemic failure analysis can furthermore be used to promote customer/supplier relations and as a basis for court opinions but also out-of-court assessments and arbitration.

Before the procedure for a systemic damage analysis is discussed below on the basis of the individual steps, the terms failure and damage must first be described and defined. From this, the breadth of the usability of the damage analysis becomes clear at the same time.

For further literature on this chapter see [60].

2.2 Definition of terms

In failure analysis, it is important to distinguish between a failure, damage, primary damage and consequential damage. This helps a lot in structuring a damage event and in communication between other experts who are part of the failure analysis team.

Failure

A failure is defined as a clear deviation from a desired or nominal state that exceeds the limit value. Failures do not have to interfere with the basic function. As an example, one can mention failures in the printing of packaging materials. The essential requirements, such as fixing the contents or protecting the contents from external influences, remain in place, but the failure causes a restriction in the appearance of the packaging or product. In this respect, this failure has a technical cause, but is not a technical damage, but it can lead to an economical damage.

Damage

Damage is deemed to have occurred, if the product exhibits deviations from the intended state, which significantly impair the intended functionalities or make them impossible. In the packaging example above, this is the case if the printed image on the packaging is so poor that important information is no longer legible. The informative functionality of the packaging is thus severely restricted.

The boundaries between the terms failure and damage are not always clear-cut. For example, the supplier of a product can quite rightly define a (merely) decorative restriction in the printed image as damage because in many sales situations the product can be presented to the customer exclusively via the packaging (e.g. toothpaste tubes). It must be expected that the customer standing in front of such a product will inevitably infer the quality of the contents and will therefore tend to choose a decoratively unrestricted product. An economic loss due to decorative deviations is therefore to be expected.

The extent to which the distinctions between the terms failure and damage sometimes require detailed consideration becomes clear when the example of surface failures caused by poor printing is transferred to the automotive outer skin application. Here, even the smallest deviations, which can only be detected by trained experts and under certain lighting conditions, are interpreted as serious damage because the integrity of the painted surface is one of the most important requirements in the automotive sector.

Primary damage, consequential damage

In the evaluation of damage cases, the distinction between causal primary damage and resulting consequential damage is of great importance. Consequently, the cause of a consequential damage is a primary damage. Often, primary damage immediately shows only a minor technical functional restriction, whereas consequential damage has dramatic effects that can completely mask the existence of primary damage.

A typical example of a chain of consequential damage is the incorrect processing of plastics. Processing at too low temperatures can lead to a critical temperature increase of the melt due to friction, which can further cause thermal degradation of the polymer. In use, such products sometimes exhibit significantly more brittle behaviour coupled with reduced load-bearing capacity. If such products fail, the cause may be found to be a reduced average molar mass – provided suitable reference samples are available – and this may be attributed to the granules. In this case, however, the reduced molar mass is not the primary cause, but a consequence of processing under unsuitable conditions.

Reference samples

As will be shown below, the defective product represents a very important information carrier in the sense of a "corpus delicti". Some information can be "read out" by instrumental analyses (see below). However, many of the results from instrumental analyses cannot be interpreted on their own. Design bases and drawings as well as material data sheets are needed, which are not always available. Therefore, reference products are very helpful. Reference products can often be obtained from reserve samples.

However, the term reference is interpreted very differently. Generally, it is understood to mean a product in a target state. However, a reference sample can also be representative of a design, a tool sample, a batch of raw materials, a production batch, etc. A reference sample can either be a new part without any load history or a part from field use or test run with inconspicuous result. The significance of analysis results depends in very many cases on the existence, authenticity and representativeness of a reference sample. Against the background of the parties involved, it must therefore always be critically questioned what the reference sample in question represents a reference for and who is primarily responsible for assigning it this role in the context of the damage analysis. It must be checked whether the independence of the investigations can be preserved at all by taking into account the reference sample defined and presented as such, or whether this could possibly already lead indirectly to taking sides.

2.3 Carrying out failure analyses

The following explanations describe the procedure for performing damage analyses. Figure 2.2 shows the steps of the failure analysis schematically, which will be elaborated on more concretely in the further course. The procedure is to be understood as a kind of guideline, which points out the most important aspects to the user, but does

Figure 2.2: Stages of the failure analysis process.

not relieve him of his responsibility since the procedure must be individually adapted in each case. Following the procedure, some examples are outlined.

Similar to an ordinary project, failure analyses always take place in a triangle of tension of the benefits under costs and time pressure (Figure 2.3). The balance can vary depending on the client: In case of an expert opinion ordered by a court the effort for the analysis has to be estimated and the costs are more or less fixed before the analysis starts. The court defines a deadline whose compliance is to be checked by the damage analyst. In the evidence warrant drawn up by the court, the objective, namely benefit, is defined. For out-of-court failure analyses, the balance looks somewhat different in most cases. Here, time pressure plays the most important role, as products manufactured due to damage cannot be brought to market or entire production lines can come to a standstill. In addition, the time it takes to resolve the problem is harmful to the image of the client or other interested parties.

In addition, in failure analysis there is the sphere of influence of information, which can be a very sensitive issue because access to information becomes subject to the strategic actions of the parties. Chapter 2.3.2 gives a more detailed view to this aspect.

2.3.1 Capture and documentation of the damage pattern and the environment

At the beginning of a damage analysis is the comprehensive assessment of the damage pattern itself. The object should be personally inspected in the environment where the damage was originally detected because it allows all influencing factors to be recorded. This is why practical failure analysis should not just happen at the desk. It starts at the scene, on the fully automated production line of food packaging, in the

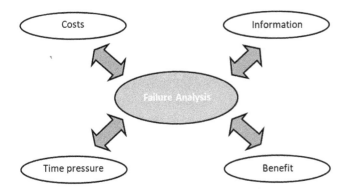

Figure 2.3: Failure analysis under the influence of interested parties.

trench of an exposed pipe burst or in the basket truck for the inspection of defective roof elements. The sequence and framework conditions could not be more diverse. Since damage is often directly related to costs and personnel responsibility, such on-site visits often take place with a number of interested parties. Claimants and causers, lawyers and supposed witnesses all do their very best to influence the course of the subsequent investigation at this early stage by giving their opinions. It is sometimes not easy but all the more important for the damage analyst to maintain complete objectivity in these emotionally charged events. These appointments can be quite uncomfortable because the analyst finds himself directly between the lines.

The damage image is generally recorded with the naked eye, a magnifying glass or a macroscope. Preparations should be avoided in this phase, as they alter the damage pattern. If, however, it is necessary to carry out altering measures at the time of damage assessment, the initial condition and all altering steps must be documented photographically. In many situations, however, this is difficult to implement, for example, on construction sites in places that are difficult to access, under adverse conditions or where samples must be taken with the help of craftsmen from corresponding trades. Furthermore, one does not always have the opportunity to be personally present at the moment of sampling and to be able to document the exact sampling position and procedure oneself. The following points should therefore at least be taken into account as best as possible when recording the damage:

- Installation situation on site (taking into account applicable guidelines, e.g. VDE or DVS)
- Interaction with other products or components within given assemblies
- Operating conditions on site (e.g. temperature, weather and human factor)
- General functionality of the defective product
- Information on possible further processing steps (e.g. welding records)
- Recording of time, locality version and, if necessary, cavity information
- Manufacturing characteristics of the product
- Information about the material, if possible inclusion of data sheets

- Any anomalies regarding the component, mould or machinery design
- Deviations from the target condition and target functionality by obviously recognisable features or by comparison with reference products.

The capture process must always meet the requirement of completeness since it must be assumed that the damage image is subject to further possibly falsifying influences in the course of time and many of the subsequent analyses have a changing or destructive character. It is therefore an inventory and securing of evidence in the forensic sense.

In series production, increased attention is usually paid to defects or damage to components when there is a failure rate classified as critical according to a given definition. This can occur linearly, exponentially, cyclic or even abruptly, as shown in the example of Figure 2.4 (left diagram). Changes in a failure quote that rises at time t_0 from approx. 2% to approx. 30% should be of an alarming character and force quick action.

Due to such a discontinuous increase in failure quote, one is initially inclined to look for an equally discontinuously changing influencing variable around time t_0. However, this search often remains fruitless. The reason for this is that even slowly and continuously (creeping) changing influencing variables can have an undesirable effect on failure quotes as shown in Figure 2.4 (right diagram). The quality relevant product property does not vary significantly before t_0 and after t_0, but there is a continuous decrease over a long period, even before t_0. The search for influencing factors is particularly difficult in these cases because no significant effects will be identified in the time around t_0.

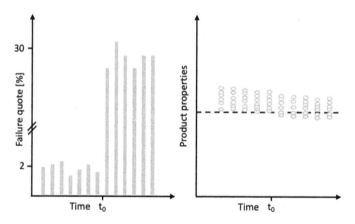

Figure 2.4: Correlation between failure quotes and "real" product properties.

2.3.2 Capture of the damage environment

As far as possible after the occurrence or the detection of damage, the damage environment should initially be recorded without further evaluation. This is to ensure that areas of influence are not disregarded too early, which, for example, could only become more significant for the development of damage in combination with other influences.

The systematic capturing of the damage environment is comparable to the situation of a medical anamnesis. Here, all information that may have a connection to the damage pattern is initially collected in a value-free manner. Such information often must be actively requested by the persons involved in damage assessment, product development, production, assembling or quality assurance.

At this point, the above-mentioned on-site meetings have a special significance. In the best case, they offer the unique and often irretrievable chance to capture the authentic, undistorted damage in real surroundings and with all the given environmental influences, to get answers to essential questions: In which environment was the damage discovered? Did the damage also occur there? Which influences exist at the place of discovery (sun, water, temperature, media, etc.)?

Unfortunately, on-site meetings are not always possible because, for economic and production reasons, the plant is often put back into operation before a comprehensive determination of the cause of the damage has taken place. In this case, photos and/or sketches should be made before commissioning.

The term *damage environment* not only means the local conditions under which the damage was perceived but also includes the entire product history from the definition of the product requirements to the damage assessment. This includes information on

– the loads and stresses to which the product is subjected
– the design criteria
– information on manufacturing (machines, parameters, etc.)
– product and material data sheets
– the history of the product in relation to the product life cycle and the product series.

Figure 2.5 shows an overview of the influences on the occurrence of damages. Starting with the definition of requirements, the assessment of the damage environment covers the product development phase (i.e. the definition of requirements) up to stresses in the use phase including all phases in between (e.g. processing).

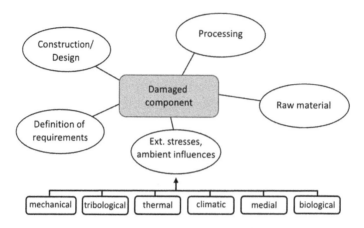

Figure 2.5: Possible influences on the development of damage.

Definition of requirements

This aspect is important in order to be able to evaluate whether all important characteristics have been taken into account in these requirements so that the product can
meet the intended stresses. Here, depending on the industry and product, a large handling variance can be observed. Within the range from applications without any existing requirement lists to the other extreme, where a requirement profile is not only
aligned with the expected regular or typical application, but any extreme situations
with excessive stress conditions are also taken up. An example of the latter case is the
automotive industry, where moulded plastic parts generally have to function permanently in a temperature range from −40 °C to +80 °C, or where components still have
to perform adequately at the material load limit. Far from these preceding considerations, the situation may have arisen here that the product's area of application has
been modified or extended over its period of use without the existing requirements
profile being re-examined and re-checked for validity.

Example Chapter 5.10: Failure of high voltage sleeves

Construction/design

Next, it must be checked whether these requirements have been correctly implemented in the design and layout and whether, under these circumstances, the material has basically been correctly selected. In the practice of failure and damage
analysis of plastic components, it is not uncommon to find that a material selection
has been made too monocausal, e.g. mainly from a mechanical point of view or litera-
ture-based resistance towards chemical attack, without sufficiently considering the

limits of process ability for the intended moulded part geometry (viscosity, post-shrinkage, shear-induced phase separation, etc.).

Example Chapter 5.21: Fatigue strength of safety-relevant plastic components made of PA6 GF30 Chapter 5.14: Leaks in swimming pool elements

Raw material

In this context, the term "raw material" refers to the overall system of base polymer(s), fillers and reinforcing materials and any additives such as stabilisers or plasticisers. Further influences can be introduced in unnoticed (material mix-up) or undocumented changes in the material composition or batch-dependent property fluctuations. Quality losses of the raw material due to storage (e.g. UV radiation, plasticiser loss or hydrolysis) can also play a role here.

Example Chapter 5.3: Cracks and delamination Chapter 5.10: Failure of high voltage sleeves

Processing

The actual processing and the associated processing errors are still the most frequent cause of failure in plastic components, which is why this area of influence is of great importance. The most common processing defects that result in an immediate risk of subsequent component failure include excessive internal stress or orientation states, cold edge layers or weld lines and shrinkage cavities or micro crack formation, to name but a few. It has to be considered that polymers underlie ageing processes already during processing, predominantly due to heat and shearing. However, one should not be tempted to make sweeping generalisations in the effect evaluation and prematurely define the mere existence of the defect as the inevitable cause of damage. Processing defects, their manifestation in the moulded part and the resulting risk of potential component failure must be evaluated on a material-specific basis, which requires appropriate experience. With this perspective, it also becomes clear that the real quality of a plastic part is always a compromise between perfect process parameters and economical considerations. The present examples are the mould and melt temperatures: Within a reasonable frame higher values (especially in term of melt temperature) will reduce orientations and inner stresses, but this will increase the costs because of the longer cooling time. These interrelationships require a balanced choice of process parameters.

Depending on the defective components to be investigated, the field of influence "processing" must be extended to further processing steps such as assembly operations, thermoforming, joining (welding, gluing, etc.), coating (calendering, painting, CVD, PVD, etc.), thermal storage, etc. The combination of processing steps makes failure analysis more complicated, as failures or errors often occur after the last process-ing step, but the cause may lie in an earlier processing step. As an example, adhesion problems of galvanic coatings on injection-moulded plastics parts can have the cause in the galvanic process as well as in the injection-moulding process:

– The necessary mechanical anchoring of the metal-layer through undercutting is not given due to too weak or too strong staining.
– Due to strong shearing of the melt during mould filling the different phases of the plastics compound are sheared or even destroyed. This makes it impossible to carry out a galvanic process under the usual conditions.

In conclusion, it must be said that the processing of plastics has many sensitive effects on the properties and behaviour of plastic parts. The knowledge of the fundamental interrelationships and the appropriate analysis methods are key elements of a sys-temic damage analysis.

Example Chapter 5.9: cracking of chain links, Chapter 5.7: Blistering on electronic con-nectors, Chapter 5.11: Cracking of toilet cisterns, Chapter 5.13: Polyoxymethylene rack breakage, Chapter 5.21: Fatigue strength of safety-relevant plastic components made of PA6 GF30

External stresses/ambient influences

One of the first thoughts that come up in the event of damages is the misuse of the product. This means that there must have been environmental influences in the prod-uct's life, the extent and characteristics of which were beyond the defined requirements and thus probably not taken into account in the product design. In this case, all possible ambient influences have to be taken into account and have to be assessed. This is a quite complex and extensive task. In order to structure this procedure, we distinguish the following most important environmental influences based on the frequency of their occurrence, following the guideline 3822 "Damage Analysis" of the VDI [N.N.21]:

– *Mechanical:* All mechanical loads lead to deformation, fatigue and/or fracture.
– *Tribological:* Stressing of the surface of a solid plastics body through contact and relative movement of a solid, liquid or gaseous counter-body.
– *Thermal:* Thermal stresses can lead to deformation or easier deformability, ther-mal or thermo-oxidative degradation.
– *Climatic:* Weathering stresses that can be caused by the effects of irradiation (UV, VIS, IR), changes in humidity and moisture as well as temperature changes. This applies in particular to the use of plastics outdoors.
– *Medial:* Stresses due to solid, liquid or gaseous media, especially with additional inner and/or outer mechanical stresses.
– *Biological:* Damage caused by microorganisms as the sole or accelerating cause.

The effects of these influences on plastics are very divers. Some plastics are quite resistant to some of these influences, and others are not. Therefore, in the damage acquisition phase, it is always important to evaluate critical influences against the background of material-specific sensitivities or reactions. In many cases, this consideration reduces the number of critical influences and focuses only on a few.

Decades of practical experience in damage analysis show that almost all cases of damage can be well categorised on the basis of the five influences mentioned above. An exception to this is, however, the electrical influence, the effect of which can generally be a thermal stress up to thermal degradation. Although, the damage patterns are so striking that this effect is not further differentiated here.

It is, of course, a simplistic model to assume that a plastic part is only under the effect of one of these influences in its lifetime. Nevertheless, when capturing the damage, it is very helpful to first consider the possible influences as individual loads and only then to take into account possible interactions or synergy effects on the plastic component. In the course of the damage capture, all influences are to be described in principle; in the course of the development of the damage hypotheses (see below), they are evaluated in more detail in connection with other influences such as "material".

It should also be noted that the lifetime of the plastic product starts immediately after processing. Therefore, storage conditions must sometimes also be taken into account, as ageing of plastics under insufficient storing conditions can already lead to change of properties. However, it must be taken into account that ageing of the polymer itself starts with storing, drying and processing.

The diagram in Figure 2.5 can basically be used as a model for capture of the damage environment as well as later for developing the damage hypotheses. Adaptation and detailing are always necessary with regard to the case under consideration.

Example Chapter 5.18: Cracks on circulations pump housings, Chapter 5.2: Component failure: Justified complaint?

For further literature on this chapter see [10, 11].

Gathering information in failure analysis

More parties – more complexity: The need for failure analysis can arise in very different environments. Especially in the case of damage analyses with a high economic value and several parties involved, it must be taken into account that there are different information holders who have different expectations of the outcome of the damage analysis. Obviously, this is the case when the failure analysis has been commissioned in a judicial context as a legal expertise.

Blind spots, availability: Much of this information is already provided by the parties involved in existing damage cases. However, it is precisely this phase of a damage analysis that requires active inquiry in order to be able to illuminate the "blind spots" of the parties involved. In practice, however, it often turns out that the desired product history information (e.g. design criteria and material release) is not available or not provided. The failure analyst must record which information was provided by which party, as knowledge of the origin of information can take on a new meaning at a later stage of the failure analysis, with the help of which the authenticity of the information can be assessed against the background of the possible protection of interests of individuals or groups.

Figure 2.6 illustrates a typical situation which shows that the client, as the one who has a vested interest in finding the causes of defects, often does not have direct access to all the information about the entire life of a product. At the same time, it becomes clear that in a case of damage, the information he receives from his supplier or his customer may be shaped by their interests.

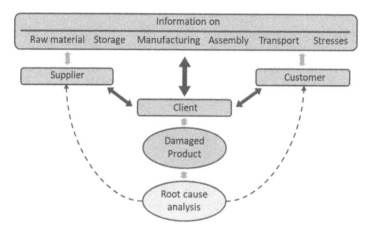

Figure 2.6: Indirect access to necessary information.

Information management: Consequently, high demands are placed on information management. Figure 2.7 shows schematically that, especially in the context of complex damage analyses with several parties involved, the damage analyst has a great need for information (1, 2, 3, 4) on the one hand, but on the other hand is also provided with a wealth of information, the significance of which is not foreseeable for the damage analyst at this stage. For example, information (6) is deliberately offered that supports the damage hypotheses of an involved party but has no direct correlation to the failure, without this being recognisable to the damage analyst at the time. And of course, there will be a lot of information requested by the failure analyst, which is requested, but not offered but important (3) or requested and offered, but not important (5). Furthermore, there will be information that is requested, but is not offered and is not important.

As the information cannot and should not be assessed at this stage, all this information and all this missing information needs to be managed and documented. Some of this information can be classified, assessed and used at a later stage. It will always be important at that time to be able to access the timing and source of the information.

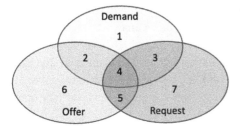

1. neither offered nor requested, but important
2. offered but not asked for, nevertheless important
3. requested but not offered, but important
4. offered and inquired and important
5. offered and demanded, but not necessary
6. offered, not requested and also not necessary
7. Request but not offered and not necessary

Figure 2.7: Quotation of important and unimportant information.

Classification of serial products: A special feature of plastic products is the fact that they are usually mass-produced. Failure analysis can be greatly simplified and accelerated if the damage is classified concerning this.

– Individual damage (I): Individual damage of type I is characterised by the fact that equivalent products are already on the market, and the damage in question represents an individual case according to the information available. This can significantly simplify the damage analysis, as the product as such can be considered, tried and tested. For this reason, it is initially less likely that mistakes were made in the definition of the requirements, in the selection of materials, in the design or in the processing. The first investigations should therefore focus on the use phase (see Figure 2.8).

However, it must be critically examined whether the case of damage is actually an individual case by making a comparison between defective (not o.k., "n.o.k.") and non-defective ("o.k.") samples. Here, among the o.k. samples, those with a history as similar as possible to that of the n.o.k. samples are to be preferred. It should also be borne in mind that classification in this class can only be a snapshot, namely if further cases of damage become known in the course of the investigations. Then these damages are to be classified as individual damages (II).

– Individual damage (II): Individual damage of type II differs from type I in that defective products occur frequently, but not all mass-produced products are defective. Here, too, it can initially be assumed, as a first approximation, that the product is tried and tested since market experience already exists. The influence in which the defective products differ from the non-defective products must be examined. The investigations should therefore primarily serve the purpose of attempting to allocate production periods and material batches, etc.

An existing damage case often influences the perception threshold of the parties involved. This leads to the methods and sensitivities for quality assessment being changed after the damage has become known, even before the damage analyst is called in. The direct relationship of the methods to the damage can become blurred. It is therefore always important for the damage analyst to record the history of the quality tests.

– Collective damage: Collective damage is to be assumed if the damage in question relates to all products of a series. In this case, the damage analysis must cover all the influences shown in Figure 2.5, which can make a damage analysis very time-consuming and expensive.

In extreme cases, the individual damage of type II can turn into a collective damage if the defective products at hand originate from a very early period of series production. Then there is a risk that the products will fail prematurely only in the course of a certain operating time, which could consequently still happen to the rest of the products as well.

2.3.3 Development of damage hypotheses

Starting points for the development of damage hypotheses are the findings from the damage pattern and damage environment analyses. The general knowledge of typical weaknesses of processing procedures and known damage mechanisms of the materials has to be concerned.

In order to structure this phase, it is advisable to adapt the fields of influence (Figure 2.5) to the specific damage case, in which, on the one hand, fields of influence can be excluded by comparison with the damage environment and, on the other hand, individual fields of influence can be detailed. As a rule, it is helpful to classify the damage as shown in Figure 2.8 since this often allows complete fields of influences to be separated out. This procedure reduces complexity without the risk of losing sight of important fields of influence.

Applied to the example of misprinted toothpaste tubes (see Section 2.2), the complexity can be reduced by the fact that the product as such is established and, as a result, the design, material selection, manufacturing process, coating technology and coating system (including pre-treatment) do not have to be fundamentally and primarily questioned. The entire field of influences due to operational stresses is also excluded for the time being. Rather, it is a question of whether deviations from the target or previous conditions can be detected in these areas of influence. In particular, it should be investigated here whether the defect pattern can be assigned to specific batches, as this is an individual damage of type II (see Figure 2.8).

In this case of damage, this leads directly to the question of reference samples from earlier production since comparative investigations usually permit more meaningful results in the instrumental analyses subsequently carried out. Therefore, the

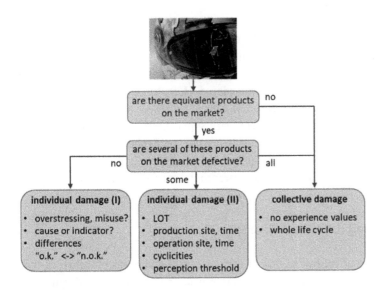

Figure 2.8: Classification of serial products.

importance of reference samples in damage analysis cannot be overestimated. Failure analysis deals with the influences and their effects not only on material properties but also especially on product properties that are typically not well known. For this reason, references are used to define the target state more precisely, whereby, as described above, it must be checked very carefully what a product designated as a reference is actually a reference for. For example, reference samples may well originate from different raw material or production batches than the damaged parts, which can make the representation of the target state blurred.

Methods for the development of damage hypothesis

In this phase of the failure analysis, it is very helpful to use additional trouble shooting methods as the creation of mind maps or Ishikawa diagrams or Six-Sigma methods. This makes it much easier to communicate the current status of hypothesis generation and validation within the team and with the client and other stakeholders.

In order to determine the multitude of complex information and to organise it accordingly, there are numerous tools and methods that can support and help with this procedure. These problem solution methods are to be assigned to the quality management and can be helpful both in preventive damage management and in clarifying of a damage case. Some well-known tools and methods are for example the Failure Mode and Effect Analysis (FMEA), cause and effect diagrams (also known as Ishikawa or Fishbone diagram) and Six-Sigma (DMAIC method).

Mind maps are very suitable for an initial collection of ideas and a first structuring. In contrast to some other methods, the observer is not guided by predefined categories or keywords and is completely free in his idea generation. The freedom of idea generation can be seen on the one side as an advantage but also as a disadvantage that possibly important aspects are not considered. Here, thoughts can be guided by the scheme of influencing factors (Figure 2.5). Nevertheless, the method is a good way to organise first ideas. With the help of software support, very comprehensive mind maps can be created that are dynamic and can accompany hypothesis generation.

In the following, some useful tools and methods are briefly explained as examples. For more in-depth information, please refer to the specialist literature.[2]

Cause/effect diagrams (Ishikawa/fishbone)

This diagram is a visual representation of various causes and areas of influence that ultimately affect the outcome and are therefore crucial to the final result (Figure 2.9). The diagram can also be thought of as a fishbone, with a line running from the left tail to the right head. In the head, the problem is entered, which should be formulated concisely and neutrally. Leading away from the centre line are the "scales" of the fish, which represent the fields of influence: human, material, machine, measurements, method and co-environment by default. These fields of influence can also be added to or modified as required. At the individual influence fields "bones" are drawn in, which symbolise the possible causes. The diagram does not show an exact solution to the problem and can become confusing for more complex problems.

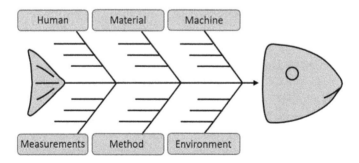

Figure 2.9: Ishikawa or "fishbone" scheme.

---

2 Gerd F. Kamiske – Carl Hanser Verlag.

Failure mode and effect analysis (FMEA)

The FMEA is actually a method to estimate a risk and a corresponding failure in advance (Figure 2.10). Here, possible negative influences and errors are evaluated with a probability of occurrence as well as their probability of detection with a key figure between 1 and 10 and then added together with various factors. The aim here is to identify processes that are particularly at risk for errors and to minimise the risk of failure or damage. It is particularly important to note that this is not a general method for damage analysis, but rather an error or damage prevention.

The FMEA can be divided into three fields of application, which are used in different areas and questions:
- **System FMEA:** In this case, individual subsystems are to be examined together in order to be able to assess their effect on the entire system. The main objective is the detection of possible errors or weak points at the interconnection points of the individual subsystems.
- **Design FMEA:** This type of analysis focuses on a single component and its design. The aim is to adapt the design in such a way that the product's probability of failure is as low as possible.
- **Process FMEA:** Based on the preceding analysis of the design, possible sources of error in the manufacturing process are considered and evaluated.

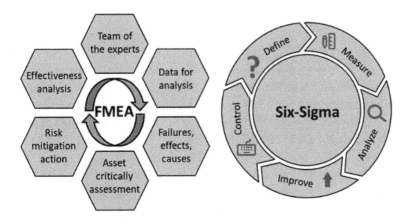

Figure 2.10: Elements of FMEA and Six-Sigma methods.

Six-Sigma (DMAIC): Six-Sigma is a process improvement concept according to which appropriately trained groups of people, so-called SixSigma Belts, with different levels of qualification, achieve quantifiable and thus directly measurable financial optimisation results. The implementation of Six-Sigma in a company starts with the comprehensive recording or identification of relevant influencing factors using statistical methods that take into account all possible input and output variables of a (faulty) process. Measurements of the actual state, the definition of target states and the achievement of these are also carried out using in-depth statistical analyses. The data basis created in this way also lays the foundation for a continuous improvement process of the company, which goes beyond the temporary occurrence of any process faults. The foresight associated with this is one of the main differences to other trouble-shooting methods, which are often only used as a means to an end in the case of suddenly occurring or sporadic errors.

The aim of Six-Sigma is to create a process that is as free of failures or lower the risk of potential failures as much as possible. The review and improvement of the core process is divided into a total of five milestones. These are clearly structured with the help of the principle "Define", "Measure", "Analyse", "Improve ", "Control" (DMAIC) and have a clear guidance for optimisation (Figure 2.10). The following questions are taken into account during the five steps:

1. Define: Start initiative – What is the failure or optimisation target?
   Documentation of the current state and determination of requirements.
2. Measure: Collect data – What are the actual process values associated with the failure?
   Quantification of the actual state on the basis of existing and newly measured significant data.
3. Analyse: Analyse data – What is the relationship between the data?
   In this step, cause-effect relationships are determined and statistically validated to identify the root cause of failures. Methods such as Ishikawa, FMEA or brainstorming can play a role here.
4. Improve: Development and implementation – What measures can be taken to eliminate the failure?
   Based on the influencing factors and causes of failures identified in the previous step, concrete improvement measures are now implemented.
5. Control: Ensure implementation – How to sustain improvement?
   Finally, the improvements achieved are evaluated against the background of the defined goals and with regard to their future security.

In most cases, these methods lead to a number of theses that harmonies with the damage pattern. Then, it makes sense to priories the theses. Criteria for the prioritisation might be their probability against the background of the damage environment, material and processing consideration. Another criterion might be effort that has to be made to test them. It can be very useful to be able to definitively exclude relatively improbable theses at the beginning by means of simple tests.

2.3.4 Determination of the test plan

An essential step in a damage analysis is the translation of the damage hypotheses into questions that can be answered with the help of instrumental analysis. In addition, when selecting the analyses, it must of course always be checked whether the sample properties are accessible to a selected analysis method. With the aid of a test plan, the tests and analyses to be specified, the chronological sequence of the tests, and the criteria for taking samples from the damage and reference samples are outlined. Likewise, non-destructive examinations will be preferred to destructive ones.

Even in cases of damage that initially appear obvious, it has proven helpful to draw up a precise test plan, as unforeseen complexity sometimes emerges after initial results. A missing test plan, especially when working in a team, can lead to the fact that it is not possible to reconstruct in retrospect the reason for which an analysis was carried out and the criteria under which the sample was taken. The latter often reduces the significance of analysis results to a critical extent. Possibilities for repeating sampling are often very limited in damage analyses.

Last but not least, the inspection plan often has to satisfy economic considerations. It therefore makes sense to weigh up whether investigations are prioritised according to the criterion of the most probable damage analysis or the least effort.

2.3.5 Carrying out instrumental analyses

Instrumental analysis is a multifaceted challenge in damage analysis. The damaged products often represent unique specimens with their individual histories. However, most of the analysis methods or the necessary preparations have a destructive character, which forces a cautious approach. On the other hand, the defective products (as well as the references) are very important information carriers that may carry traces of the damage and can be made accessible only with the help of instrumental analysis.

"Reading out" the information of a damaged part or a reference part is a particular challenge because a wealth of internal parameters determines the external characteristics. First of all, it should be noted that the material plastic consists of at least four components: the polymer (s), reinforcing materials, fillers and additives. These components in turn show their overall effect due to many properties of their own. Figure 2.11 gives a rough overview. It is known, for example, that the molecular weight and its distribution have a strong effect on the mechanical behaviour. However, additives, fillers and reinforcing materials can also have a major influence on mechanical behaviour. If, in a case of damage, the hypothesis was formed that the mechanical properties of a product had fallen short of the expected value, a whole range of detailed questions arise for analytics, each of which would have to be investigated using suitable methods. This makes the determination of suitable methods quite complex, especially since there is usually still a lack of information about the material itself at this point.

Therefore, the instrumental analyses are scheduled according to the test plan. References have to be involved in the test plan. It has to be checked to what extent the available sample quantities allow a statistical validation of the measurement results.

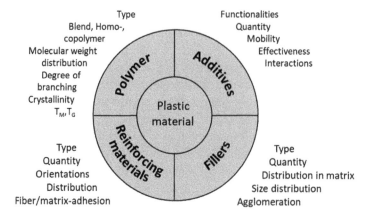

Figure 2.11: Components of plastics and some of their property-determining structure parameters.

Since analyses have to be carried out by specially trained personnel, their performance should usually be directly linked to the question generated by the damage hypotheses, as otherwise there is a risk that unsuitable analysis parameters will be chosen to answer the question. This requires the involvement of the expert personnel in the hypotheses that are pursued.

In damage analysis, the most important analysis and testing methods include, above all, light and electron microscopic methods together with the associated preparation of cross sections or thin sections. For these methods, sampling plays a special role in that it must be clear from the outset which viewing direction in relation to the component is to be examined microscopically. Furthermore, spectroscopic methods such as infrared or Raman spectroscopy, energy-dispersive X-ray spectroscopy and photoelectron spectroscopy play an important role in material identification, structure clarification, imaging of ageing states and media effects as well as in fracture pattern evaluation (fractography). Thermal analysis methods such as differential scanning analysis, thermogravimetric analysis and thermomechanical analysis help to visualise the thermodynamic state of the material (e.g. orientations, crystallinities, phase transitions, physical ageing and oxidation state). Most important methods mentioned here will be explained in Chapter 4. They have in common that their informative value generally increases significantly with the support of reference samples.

The interpretation of the analysis results is primarily carried out under consideration of the analytical questions arising from the damage theses. But also, anomalies and conspicuities in sample preparation, test performance and results that are not related to the actual problem can provide very valuable information about the cause of the damage.

In this phase of a damage analysis, a cause of damage is often considered "proven" if the results of the analysis methods are compatible with the thesis to be tested. It is easy to be carried away by the statement that the thesis has been confirmed. However, this is deceptive because scientifically the thesis is merely not refuted. However, this alone does not increase its truthfulness and says nothing about whether there is not a better damage thesis that has not been taken into account at all.

The evaluation of results of tests that are supposed to recreate the damage process is similarly deceptive. They are a proven means of testing the plausibility of a damage theory. However, it is only a question of whether a damage process can have taken a certain course and led to the same damage pattern as the actual damage. If the original damage picture cannot be generated by the adjustment test, either the thesis is wrong, or the adjustment test is unsuitable. If, on the other hand, the damage pattern is reproduced, this is not yet proof of the thesis since other theses can also lead to the same damage pattern. In this respect, damage analysis remains a non-scientific method in the narrow sense.

If interim results imply or allow new hypotheses, the test plan can be changed, whereby good documentation is to be emphasised.

2.3.6 Designation of well-founded influences and causes

The causes of the damage are presented in comparison with the damage pattern, the damage environment and the compatible theses in this respect. The course of the damage must also be described. It is recommended to describe this in a detailed report that covers the holistic course of the failure analysis, starting with the documentation of the damage pattern. Figure 2.2 gives a possible outline structure of the report.

In practice, damage is usually not monocausal. Various influences must be named for a damage pattern because each of them has an effect on the damage event. These influences from a constructional, material and processing point of view, coupled with the stresses in use, often only become critical in their totality. It also follows that weak points are usually identified at several points in the course of a damage analysis. To illustrate this, let us consider a fictitious case of damage (which, however, often occurs in practice in this or a similar way):
- Cracks are discovered on the housing of a medical device.
- The cracks are located in areas with narrow radii of curvature, e.g. corners or screw holes.
- It turns out that the device was disinfected with an alcohol.
- An analysis on thin sections using a polarisation microscope shows very high internal stresses in this area, which are also found on new casings that had not yet been in contact with alcohol. However, the new casings do not show any cracks.

This description allows the naming of several influencing factors, which are illustrated in Figure 2.12:
- First of all, it seems obvious that the plastic from which the housing was made is not resistant to the disinfectant. However, this general statement is contradicted by the fact that the housing was only damaged at prominent positions.

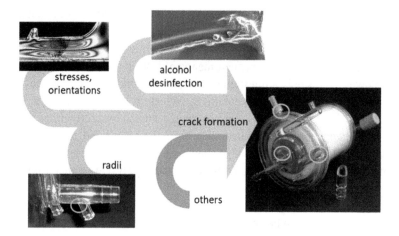

stresses, orientations

alcohol desinfection

crack formation

radii

others

Figure 2.12: Several unfavourable circumstances lead to damage.

- It can also be argued that stresses were introduced into the component during the manufacturing process. These tensions alone do not lead to damage. Only the media contact has fatal consequences.
- Furthermore, it can be stated that the constructed small radii on the housing favour the damage, on the one hand by an increased notch effect, on the other hand also by the fact that the melt was subjected to greater stress during the filling process.

This list could be extended by further issues. However, it is clear from this description that the damage could have been avoided by using different disinfectants, a different design, a less stressful injection-moulding process (e.g. higher melt temperature, possibly higher mould temperature) or a different selection of material. None of the mentioned influences seems to be able to cause the described damage in isolation.[3]

The example shows that an unsubstantiated failure analysis, which would already stop after the discovery of a first defect (e.g. the small radii), would not do justice to its task. For this reason, the damage analyst must always keep a broad view and check whether the defects found are able to interact to produce the damage. If this is not the case, further theses must be formed and investigations carried out. If this is also not clearly effective in terms of causal research, this must also be formulated as such in the report.

---

3 This description is simplified insofar as lists of requirements and specifications are drawn up for such products. In addition, the manufacturer also lists approved disinfectants that are compatible with the plastic. However, in the reality of damage analysis, the damage analyst does not have access to such information or cannot assess the reliability of the information.

2.3.7 Remedial measures and knowledge management

In the case of judicial expert reports, the work of a damage analyst is usually done with the preparation of the expert report. In the case of out-of-court failure analyses, on the other hand, the client and also other parties involved are interested in deriving further insights from the work done to determine the influences of the damage. This is because the detailed examination and evaluation of the damaged as well as the reference samples allow deep insights into product development, design, processing and material behaviour.

It is obvious to use these insights to develop remedial and improvement measures. This benefit is not limited to the defective product under consideration, but can extend to complete product groups, material groups or processing procedures within a company due to the comprehensive considerations. From this point of view, a failure analysis is designed to solve a specific problem, but in view of the effect of the findings, it also represents a quality-improving investment.

As a result, the question of how knowledge management of the findings from failure analyses can be sensibly carried out in companies arises all the more. Experience shows that failure analyses are initiated and supervised by different units or departments depending on the problem and the company structure. For example, if it is clear from the outset that the damage is related to production-related circumstances, the production management is often in charge. If internal structures have already been used to solve the problem, quality assurance is usually involved and sometimes, but not always, in charge. Furthermore, there are cases of damage that are directly managed by the development department. This shows that the findings initially run back to the individual departments. However, processes must then be developed in the company that ensure interdepartmental communication so that the greatest possible benefit can be derived.

An institutionalised "lessons learned" can provide a good framework for this. The person responsible for this must be determined by the management. In this way, the right conditions can be created so that the diverse findings from failure analysis are anchored in the right instances of a company in material selection (or release), product and tool design, manufacturing processes, product testing and all influences during product use and contribute to improvement. The approach formulated above and the guiding questions behind it provide a solid framework for this.

2.4 Final remarks

Even if the procedure shown for a systematic failure analysis appears to be time-consuming and expensive, it is ultimately only a framework that is suitable for handling even the most complex cases responsibly and without errors. If one adopts this approach and applies it consistently – even in obviously less complex cases of damage – one will find, on the one hand, that the immediately obvious is by far not always the only cause. On the other hand, one will find that the systematic approach makes it possible to focus very quickly, in that influencing factors can be deliberately faded out in order to reduce complexity without running the risk of falsification. And this makes the systematic damage analysis a technically and economically very efficient and powerful tool.

Which competences must be present in order to be able to carry out a systemic damage analysis? Figure 2.13 shows the seven skills required on the basis of the procedure described above. The essential foundation is elementary knowledge in materials technology (including the effect of external loads), in plastics processing and in product and tool design. These competences are necessary to be able to quickly classify typical and generally known mechanisms in a case of damage. This is the basis for competences in instrumental analytics, which enable the translation of the developed hypotheses into the language of analytics. This allows a rapid assessment of which method can be used to investigate questions, what kind of results are possible, what the permissible interpretation framework is and what requirements must be placed on the sample properties in order to be accessible to an analytical method.

In damage analyses, it is essential to master a systemic approach that meets all the above-mentioned requirements so that reliable results can be achieved. The ability to communicate is important for two reasons: On the one hand, damage analyses usually take place in an area of tension between several interested parties who place special demands on the procurement and evaluation of information. Asking the right questions and classifying answers requires a certain routine in order to be able to keep communication channels open for the future through respectful interaction, even in the case of obvious misinformation. On the other hand, it is obvious that the wealth of necessary competences can only be provided by one and the same person in a few cases. Damage analysis is therefore typically a team effort that requires good and clean communication within the team. The team members who carry out the instrumental analyses, for example, must be intensively informed about the underlying hypotheses without being "bombarded" with information. In addition, the interpretation of the analysis results must take place with the experts concerned. This also requires balanced communication.

Thoroughly conducted damage analyses have a great benefit in many fields. In addition to the aforementioned damage prevention, the directly associated cost reduction, product improvement and feedback into product and process development, damage analyses have a positive effect on customer and supplier relationships. The latter will first realise through damage analyses that great attention is paid to product quality. Second, the supplier will be told in great detail which quality criteria the products must fulfil in the future. This encourages the supplier with regard to his product quality.

Material

Construction

Analytics

FOCUS OF
FAILURE
ANALYSIS

Processing

Communication

Systematics

Ambient stresses

Figure 2.13: Competences required to carry out systemic damage analyses.

It also turns out that the relationship with customers is also very well promoted by holistic damage analyses. Although in the first approach there is a tendency to avoid the
word "damage" towards the customer as a matter of principle; in the second approach
it becomes clear that the customer learns to appreciate when errors that occur are
dealt with duly. It helps the relationship with the customer if the utilisation of systemic
damage analyses proves that continuous product and process improvements are being
worked on actively and not just reactively. In this way, failure analysis can be excel-
lently integrated into existing quality standards such as ISO 9001:2015 as a multifaceted tool.

After this chapter has presented the basis of the damage analysis procedure, the
next chapter briefly introduces basic information concerning plastic material and processing technology and the most common and frequently used instrumental analysis
methods in the context of damage analyses. The distinction between these chapters
should make it clear that failure analysis and instrumental analysis are completely different. In the following chapter, methods of computer-based simulation are examined
with regard to their increasing importance as an analysis tool for failure analysis. Finally, examples from the practice of damage analysis follow, which are presented
streamlined according to the presented approach.

Rainer Dahlmann, Tobias Conen

3 Short overview of plastics

Failure analysis of plastic products requires a deep understanding of plastics materials science as well as plastics processing technology. This understanding cannot and should not be provided within the scope of this book. This chapter merely gives a brief overview of plastics and plastics processing and provides references to further reading. The chemical background, properties and conditions of various plastics are touched upon; the main focus is on the resulting possible failure patterns. The same applies to further processing in the manufacturing processes of the plastics industry.

3.1 Introduction

Polymers consist of macromolecular chemical molecules that are mostly based on an organic structure of mainly carbon and hydrogen atoms. They are natural or synthetic materials made up of a large number of monomers. In nature, macromolecular structures have multiple applications such as DNA, cellulose, starch and resin. In technical applications, polymers are known in particular as plastic materials in e.g. components, as fibres or as additives such as adhesives or coatings.

3.2 General properties and applications

Plastics are characterised by their long lifetime, high chemical resistance and their low density in the range of 0.8–2.2 g/cm^3. This is a clear advantage compared to aluminium (2.7 g/cm^3) and steel (7.8 g/cm^3). The mechanical properties can be adjusted over a wide range. Plastics can be hard, stiff and brittle but also soft, highly elastic and ductile. In addition, other properties such as colour can be changed by adding other substances. A deeper insight into the characteristics of plastics is provided by the book "Menges Werkstoffkunde Kunststoffe" or "Material Science of Polymers for Engineers" [1, 59].

Figure 3.1 shows that in 2021, 83.1% of plastics produced worldwide were thermoplastics and only 7.1% were other types of plastics. In addition, 9.8% of the plastics were reused or based on a biological starting material. Thermoplastics can be further subdivided into the standard thermoplastics PE, PP, PS and PVC and the technical thermoplastics. The standard thermoplastics account for the largest quantity of plastics with a production share of 64.4%.

Figure 3.2 shows the use of plastics over different markets. The packaging industry in particular is characterised by short-life plastic articles and thus a high con-

https://doi.org/10.1515/9783110785647-003

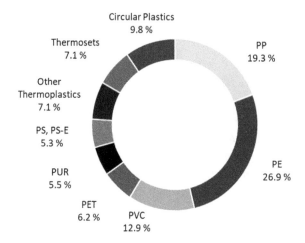

Figure 3.1: Percentage of production volumes worldwide in 2021 (adapted according to [45]).

sumption of plastic material. Both standard and technical plastics are used in automotive and electronics. These sectors processed 8% and 7%, respectively, in 2021. Other sectors with a high use of plastics include agriculture and household products (including sports, games and leisure).

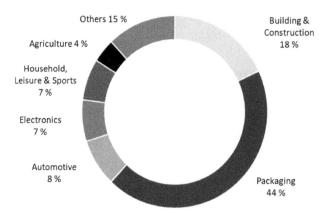

Figure 3.2: Percentage of plastics processed by industry worldwide in 2021 (adapted according to [45]).

Due to their great importance and market share, the following deals in particular with thermoplastics. These are used in all industries and are also affected for the majority of the damage cases that occur. Nevertheless, elastomers and thermosets will also be mentioned in the book.

3.3 Standard and technical plastics

Thermoplastics are often divided into standard and technical thermoplastics. Compared to standard thermoplastics, technical thermoplastics have better mechanical properties such as higher strength, stiffness and impact resistance. Technical thermoplastics can often be used at low temperatures or at higher temperatures. However, the distinction between technical and standard plastics is not always clear and sometimes there is a seamless transition. An overview of different standard and technical thermoplastics is shown in Figure 3.3.

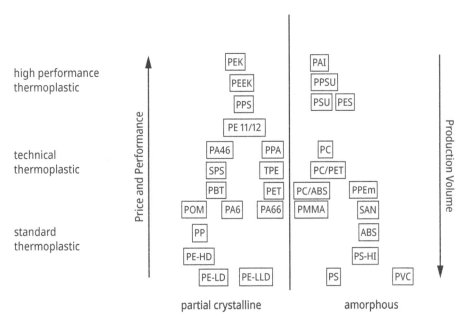

Figure 3.3: Standard and technical thermoplastics (adapted according to [1]).

Table 3.1 provides an overview of selected properties of standard thermoplastics. Table 3.2 lists the selected properties for technical thermoplastics.

Of course, this gain in application temperatures is also expensive in terms of cost. Standard thermoplastics usually cost between €1 and €2 per kilogram of granulate [57]. The technical thermoplastics range in price from €2.50 to €4 per kilogram of granulate [57]. This can be attributed in particular to the different raw materials and thus to the production of the plastics. Standard thermoplastics require simple monomers, which are often available as a waste product of petroleum chemistry or can be easily produced from them. The technical thermoplastics are produced from more complex monomers and the polymerisation processes are also often more complex.

Table 3.1: Standard thermoplastics and selected properties (adapted according to [55, 56]).

Plastic	Operating temperature (°C)	Young's modulus (MPa)	Tensile strength (MPa)
Polypropylene (PP)	0–100	1,000–1,800	~35
Polyethylene high density (PE-HD)	Up to 80	~1,000	20–30
Polyethylene low density (PE-LD)	To 80	~200	8–10
Polystyrol (PS)	To 70	~3,200	~55
Polyvinylchloride (PVC)	−50 to 60	1,000–3,500	50–75

Table 3.2: Technical thermoplastics and selected properties (adapted according to [55, 56]).

Plastic	Operating temperature (°C)	Young's modulus (MPa)	Tensile strength (MPa)
Polyamide 6 (PA 6)	Up to 90	~3,000	~80
Polyethylenterephthalate (PET)	−50 to 100	2,800–4,500	~80
Polyoxymethylene (POM)	−50 to 110	~3,200	~65
Polybuthylenterephthalate (PBT)	−50 to 140	1,700–2,700	~60
Polymethylmethacrylate (PMMA)	Up to 90	2,700–3,200	~70
Polycarbonate (PC)	Up to 125	~2,400	~65

3.4 Polymer structure

The resulting properties of plastics are significantly influenced by their structure. The properties can be attributed to the individual chemical bonds in the molecule. Figure 3.4 shows the relationships between the properties of the polymers and the various bonds in the polymer, which can be classified into primary bonds, secondary bonds and crystallite formations. The primary bond is the direct covalent bond between two atoms. The binding energy of the primary bonds is in the range 40–800 kJ/mol [21]. The main chain, side chains, substituents and the chain ends are formed from these bonds. The primary bonds are predetermined by the synthesis of the polymer and can only be changed chemically. The primary chemical shape of a polymer chain determines its type of molecule, the molecular shape, the size of the molecule, the molecular order and the incorporation of foreign molecules. Accordingly, this includes different tacticities (syndiotactic, isotactic and atactic) of substituents, cis-trans isomers of incorporated double bonds or copolymerisations and their sequence (random copolymers, alternating copolymers, gradient copolymer, block copolymer or graft copolymer) as well as branching and crosslinking between different polymer chains. Cross-links in particular are added during the processing of polymers in the downstream stage. This is often referred to as vulcanisation or cross-linking reactions. This chemically changes the polymer structure and thus the primary bonds, which is why the properties of the polymers change fundamentally as a result.

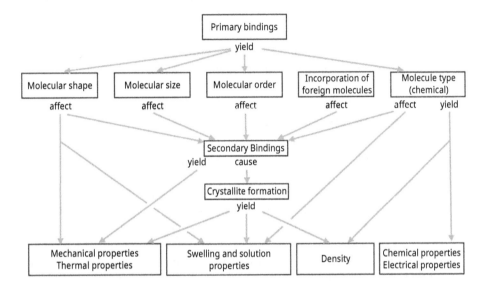

Figure 3.4: Relationships between molecular properties and material properties according to (adapted according to [1]).

Building on the primary bonds, secondary bonds can form between different polymer chains. Secondary bonds include dipole-dipole bonds, induction bonds, dispersion bonds and hydrogen bonds. The binding energy of the secondary bonds is in the order of 2–20 kJ/mol [1, 21].

Dipole-dipole bonds are those bonds that form under dipoles. A dipole is formed as soon as two atoms with clearly different electronegativities enter into a covalent bond. This is the case, for example, in a bond between fluorine and carbon. The more electronegative atom pulls the electrons of the other atom in its direction, causing the latter to become more negatively charged. The other atom, on the other hand, becomes more positively charged because the electrons are missing there. If an additional dipole now comes into contact with this dipole, the positive and negative sides of the dipole interact particularly strongly and form a dipole-dipole bond. A special case of the dipole-dipole bond is the hydrogen bond. One speaks of such a bond when a hydrogen atom is involved in the dipole formation.

If a dipole shifts the electrons of another, actually uncharged part of a polymer chain through its electrical charge, this is referred to as induction forces. This displacement of electrical charges can also occur between two uncharged polymer chains. This bond, called dispersion bond, is the most frequently occurring intermolecular interaction. It accounts for the majority of interactions between polymer chains, but has the lowest force.

These secondary forces influence the formation of crystal structures of the polymers. Not all polymers can form crystal structures. A distinction is made between amorphous polymers, without crystalline structures and semi-crystalline polymers. Semi-crystalline polymers have both crystalline and amorphous areas. The structure of the polymer, as well as the intermolecular interactions between the polymer chains, influences the resulting mechanical and thermal properties, the density, swelling and dissolving properties and the chemical and electrical properties of the plastics.

3.4.1 Polymerisation reactions

The previous chapter showed that polymers and their properties are based on the interactions between the macromolecular chains. However, the interactions that occur are influenced or predetermined by the structure of the individual macromolecular chains. The different macromolecular chains are synthesised by different (poly-)syntheses, which will be explained in more detail in this chapter.

Monomers serve as the basic structure for polymers. These are reactive molecules that allow a continuous chain reaction. The chain reaction is made possible by a reactive double bond or by at least two functional groups. The reaction of several monomers results in so-called oligomers, which have a significantly higher molecular weight. If the chain becomes even longer, we speak of polymers. Polymers are often characterised by a molecular weight of more than 10^4 g/mol. The viscosity behaviour also changes. The viscosity of non-polymer molecules increases linearly with increasing molecular weight, the viscosity of polymers increases with the molecular weight to the power of 3.4.

The (poly-)syntheses types differ in chain polymerisation or step polymerisation. In the case of chain polymerisation, a distinction can also be made between radical and living polymerisation. Step polymerisation includes polycondensation and polyaddition.

Chain polymerisation

The chain reaction is based on an unsaturated bond, such as a double bond or an unsaturated ring, which is activated by an initiator and subsequently persists as the active site in a chain until a chain termination occurs. In radical polymerisation, the unsaturated bond is activated by a radical that becomes part of the polymer chain. The radical site then transfers to the carbon atom in the monomer. This in turn can activate further unsaturated bonds on other monomers, allowing the chain to grow. The chain reaction is terminated, for example, by a combination of two active chains.

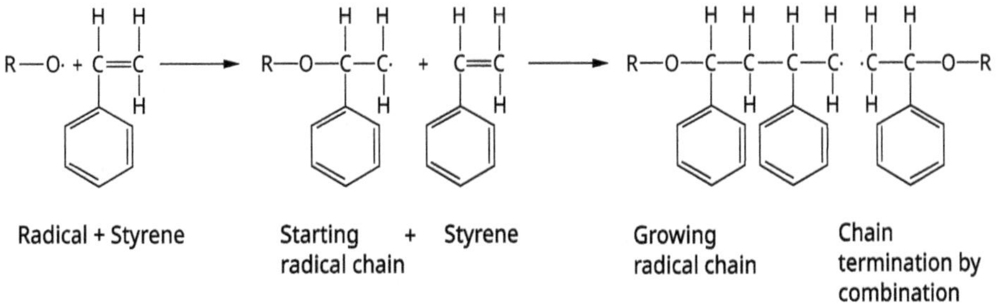

Figure 3.5: Radical polymerisation of styrene (adapted according to [1, 21]).

As an example of a radical polymerisation, the polymerisation of polystyrene is shown in Figure 3.5.

In addition to radical polymerisation, there is also living polymerisation, which can be divided into catioinic and anionic polymerisation. The essential difference is the chemical structure of the active centre. This is an anion in the first case and a cation in the second. Living polymerisation is characterised by the fact that no termination or chain transfer reactions occur. The reaction rate of the initiation reaction is much higher than that of the propagation, which is why the active chain ends are present quantitatively at the start of the reaction. Living polymerisation allows polymers with a very narrow molar mass distribution to be produced. Since there are no termination reactions, the chain ends remain active after polymerisation and can be specifically functionalised. To prevent premature termination of the chain reaction, it is essential to work under an inert gas atmosphere and with absolutely dry chemicals during living polymerisation.

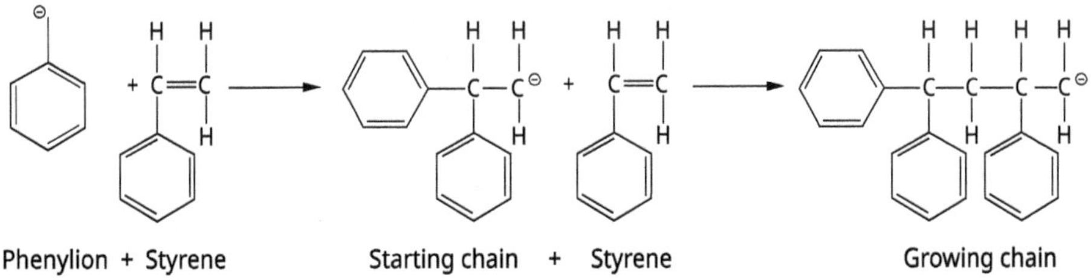

Figure 3.6: Anionic polymerisation of styrene (adapted according to [1, 21]).

Figure 3.6 shows the anionic polymerisation of styrene. The active anionic chain end remains until a termination reaction takes place. This also allows the reaction to continue if additional monomers are added.

Step polymerisation

In step polymerisation, the monomers have at least two functional groups (A and B). These can be, for example, alcohol, carboxy, isocyanate or amine groups. Two variants are possible for the monomers. On the one hand, a monomer has two different functional groups (A-B monomer) or there are two different monomers with two similar functional groups (A-A monomer and B-B monomer). For full degree of polymerisation, A-A monomers and B-B monomers must be present in a ratio of 1:1.

In polycondensation each reaction is a condensation reaction and a smaller molecule splits off. Most of the by-products have to be removed from the reaction so that the reaction equilibrium is shifted in the direction of the polymers (e.g. PET, PC and PA). Besides polycondensation, there is also polyaddition, which proceeds without the formation of by-products (e.g. PUR and epoxy).

In step polymerisation, it should be noted that the polymerisation is an equilibrium reaction. This means that the reaction can also proceed in the other direction and under the right conditions (temperature/pressure/chemicals) a reverse reaction can occur. This would result in a chain degradation which might become a relevant aspect in failure analyses. Reversal of polymer synthesis is particularly present in polycondensation, where the presence of the by-product favours the reverse reaction. For this reason, the presence of e.g. water in the processing of polycondensates can also cause chain degradation. The reverse reaction rarely occurs in chain polymerisation and polyaddition.

Failure mechanisms of different polymer classes

Different polymer classes have different failure mechanisms, which should be known when selecting the polymer. For example, the aforementioned hydrolysis leads to a reversal of the polymerisation reaction, especially for plastics such as PET, PBT, PC, PA and PU. Polymers such as PC, PE, PP, PSU, PPSU, ABS and PS have low UV resistance. This leads to discolouration, cracking and loss of mechanical strength. In PVC, hydrochloric acid decomposition leads to the formation of brown-coloured polyenes. Plastics such as PC, PMMA and PS are sensitive to stress cracking, especially in combination with special agents such as alcohols. To prevent such effects, polymers can be supplemented and improved with fillers and additives. More detailed information can be found in VDI guideline 3822 (adapted according to [60, 61]).

Molecular weight distribution

The molecular weight has a significant influence on the properties of a polymer material, as it is a measure of the chain length. The processability is strongly influenced by the molecular weight, as the viscosity increases strongly with increasing molecular weight. With longer chains, more entanglements also form between the polymer chains, which increase the impact strength and make the polymers more resistant to degradation reactions. A high molecular weight often improves the properties of the polymers, but makes them more difficult to process due to their high viscosity.

Due to the polymerisation processes, polymers do not have a uniform molecular weight. There are many different chains of different lengths with different molecular weights. Thus, a molecular weight distribution is formed, which depends in particular on the polymerisation process. Ionic or living polymerisation forms the narrowest molecular weight distribution, followed by step polymerisation. Radical polymerisation has the broadest distribution.

An example of a molecular weight distribution is shown in Figure 3.7. It can be seen that the molecular weight of the polymer chains is distributed over a more or less distinct range. The number-average molecular weight $\bar{M}_n$, the weight-average molecular weight $\bar{M}_w$ and the viscosity-average molecular weight $\bar{M}_v$ are usually given as average values.

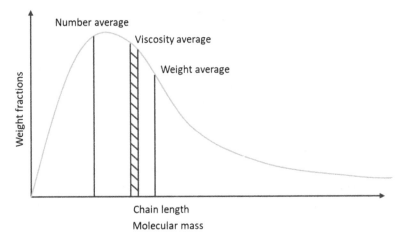

Figure 3.7: Molecular weight distribution of a polymer with the number-average, viscosity-average and weight-average molecular weight (adapted according to [21]).

The width of the molecular weight distribution indicates how many chains with different chain lengths are present. In general, it can be assumed that a narrow molecular weight distribution results in more uniform characteristic values. For example, the thermal softening range is narrower and there is less sensitivity to stress cracking and better chemical resistance. If there is a broad distribution of molecular weight, this offers advantages in processing, as the shorter fractions have a softening effect. This also leads to a decrease in brittleness.

A characteristic value for the width of the molecular weight distribution is provided by the polydispersity index (PDI). The PDI is defined according to the ratio of weight-average molecular weight to number-average molecular weight (eq. (3.1)):

Equation 3.1: Polydispersity index

$$\text{PDI} = \frac{\bar{M}_{\text{w}}}{\bar{M}_{\text{n}}}$$

As already described above, living polymerisation in particular achieves a PDI close to 1. Polycondensations are characterised by PDI values at 2 since the molecular weight is often doubled at the end of the polymerisation reaction by the combination of two chains.

For further literature on this chapter see [14, 4].

3.4.2 Crystallisation

Polymers and especially thermoplastics can solidify amorphous or semi-crystalline. In an amorphous polymer there is no long-distance order, and the polymer chains are unstructured and entangled. In the case of semi-crystalline polymers, in addition to the amorphous areas, there are also crystalline areas in which there is a regular periodic arrangement of the chains. A completely crystallised polymer does not exist since the different chain lengths, defects in the main chain, secondary chains, chain ends as well as entanglements between different chains prevent complete crystallisation.

Depending on the boundary conditions during crystallisation, two different types of polymer crystals can form. With a relaxed melt and a low velocity gradient, lamellar crystals form (Figure 3.8). With a high degree of orientation and subsequent heat-setting, needle crystals can form in PE-UHMW, for example. These are used, in particular, in fibres.

A lamellar morphology is formed in which several lamellar crystals alternate with amorphous areas. This lamellar morphology can subsequently form three-dimensionally into spherulite crystal structures (Figure 3.9).

Crystal structures form when cooling from the melt within the range between the crystallisation temperature and the glass transition temperature. At temperatures close under the melting temperature, high crystal growth rates prevail; at temperatures close to the glass transition temperature, many nuclei are formed which trigger crystal growth. This has the consequence that there is an optimum in the crystallisation rate. When heating up, the glass transition temperature is the temperature at which the amorphous regions of the polymers soften, which means that the polymer chains in the amorphous region gain mobility. When the melting temperature is exceeded, the crystal structures of the polymer disintegrate and the chains that were previously integrated into the crystal also gain mobility. Then the crystal structures of the polymer dissolve and the chains that were previously incorporated into the crystal also gain mobility. This changes the state of the polymer from solid to molten.

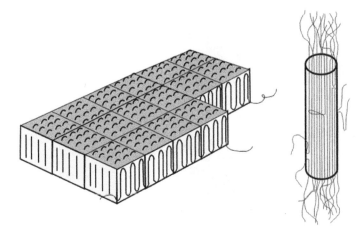

Figure 3.8: Lamellar (left) and needle crystal (adapted according to [1]).

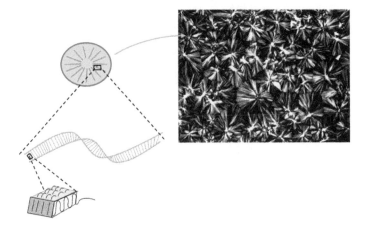

Figure 3.9: Formation of a spherulitic structure (adapted according to [1]).

Due to the different nucleation and growth rates of crystals, temperature gradients in the manufacture of products can be determined on the basis of the crystal structure. For example, many small spherulites indicate a low temperature, while larger spherulites show a slow cooling of the melt. The crystal structures also form in the direction of the higher temperature, as the growth rate is higher in that direction. Figure 3.10 shows a crystal structure formed under a temperature gradient. The crystal structure of polymers is the most compact structure in which chains can arrange themselves. For this reason, the density in this area is significantly higher. As a result, crystalline polymers shrink much more during cooling.

Due to the high packing density in the crystal, crystallisation decreases the volume. This can be seen in the pvT diagram (Figure 3.11). The pvT diagram shows the

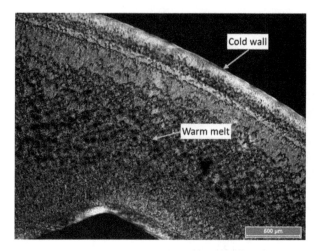

Figure 3.10: Crystal structures of a POM in a temperature gradient.

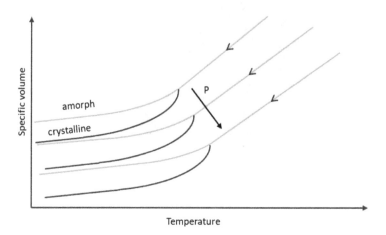

Figure 3.11: pvT diagram in yellow the amorphous cooling from the melt and in blue the crystalline cooling of a thermoplast.

specific volume versus temperature for a constant pressure. Compared to the amorphous thermoplastic, the specific volume of the semi-crystalline thermoplastic decreases significantly due to crystallisation. As a result, semi-crystalline thermoplastics exhibit higher shrinkage and warpage.

For further literature on this chapter see [7].

3.4.3 Fillers, additives and reinforcing materials

By adding fillers, additives and/or reinforcing materials, the properties of polymers can be specifically changed and adjusted to the task at hand. The material plastic is created by mixing polymer, additive and, if necessary, fillers and reinforcing materials to form the product with the desired properties. This circumstance makes the material analysis very complex, especially in the case of defect and failure analysis, since each of the components can cause a large number of defects. In addition, the interactions between the components can also lead to damage in plastic products.

The incorporation of fillers, additives and reinforcing materials is referred to as compounding and, in the case of thermoplastics, usually takes place in a twin-screw extruder with a high mixing effect. Uniform homogenisation and dispersion of the fillers in the plastic matrix is crucial for the subsequent properties. Insufficient dispersion of the fillers leads to agglomerates in the plastic, which can often be identified as a source of defects. Equally important is good adhesion between the filler and the plastic matrix, especially with reinforcing materials such as glass or carbon fibres so that the desired property improvement can be achieved.

Fillers additives and reinforcing materials can be divided into different classes, which are mostly distinguished by the desired change in properties. In particular, a distinction is made between processing aids as well as materials for the use of the plastic product. Processing aids include lubricants, release agents, anti-blocking agents or adhesion promoters. The fillers, additives and reinforcing materials for the use of the product can be divided into the following classes, for example:

- Additives that extend service life (antioxidants, light stabilisers, heat stabilisers)
- Antistatic agents
- Fillers for magnetic, electrical and thermal properties
- Flame retardants
- Colouring aggregates
- Strength-enhancing aggregates
- Stiffness-increasing aggregates
- Plasticisers
- Blowing agents
- Nano-fillers

Fillers and additives are often used in combination to achieve different properties. A distinction between fillers and additives can be made by the amount added. Often only a few wt.% of additives are added to change the properties. In the case of fillers and reinforcing materials, the proportion can be up to 60% so that the polymer is no longer the main component.

Fibre-reinforced plastics are a special group of plastics. Various fibres (very often glass and carbon fibres) are used to improve the strength and stiffness of plastics. A distinction can be made between short fibres (length below 1 mm), long fibres (length above 1 mm) and continuous fibres (length above 50 mm). Short and long fibres can be processed in extrusion and injection moulding. Attention must be paid to the orientation of the reinforcing fibres, the loss in length and the adhesion between fibre and plastic as otherwise the reinforcing effect cannot be fully exploited. Examples of this type of damage can be found in Section 5.1. A component design suitable for plastics is required for fibre-reinforced injection-moulded components.

Fillers and additives for plastic products can cause damage in certain cases, too. For example, the effect of UV stabilisers such as hindered amine light stabilisers can be consumed due to their high reactivity if the processing temperature is too high. As a result, UV stabilisation does not function and the plastic ages much faster than desired. Organic dyestuff and pigments can also cause discolouration and destruction of the pigments through oxidative or hydrolytic reactions. Inorganic fillers such as titanium dioxide or zinc oxide can become photoactive when exposed to UV radiation and, in combination with moisture or humidity, can lead to the destruction of the polymer matrix. More detailed information can be found in VDI guideline 3822 Part 2.1.3 [61].

3.5 Properties of thermoplastics

The most commonly used type of plastic is the thermoplastic. Thermoplastics are soluble in solvents and meltable when heated. They are easy to process and can be used in many different ways. They can be processed with all common processing methods such as injection-moulding, extrusion and blow moulding. Due to their easy processability, they are particularly suitable for mass applications. Additives and fillers can also be added very easily to a raw material, allowing the properties to be varied and adjusted over a wide range. The production of polymer blends can also be easily realised. The manufacturer or compounder can provide the processor with a finished granulate that can be melted and shaped into a new form with the help of temperature and shear stress. The new shape retains the manufactured product as long as it is not heated above its application temperatures.

Thermoplastics can be divided into semi-crystalline and amorphous thermoplastics. Due to the crystal structures of the semi-crystalline thermoplastics, they have different operating temperatures compared to the amorphous thermoplastics. Figure 3.12 shows the tensile strength and elongation at break versus temperature of an amorphous thermoplastic. An amorphous thermoplastic softens in the range of the glass transition temperature T_g. In this range, the polymer chains acquire more degrees of freedom and can slide along each other. As a result, the tensile strength drops significantly and the elongation at break rises sharply. This range passes directly into the flow range in which the amorphous thermoplastics can be processed. Above a certain temperature, the polymer chains decompose and the thermoplastic is permanently damaged. The typical application temperatures of amorphous thermoplastics are accordingly below

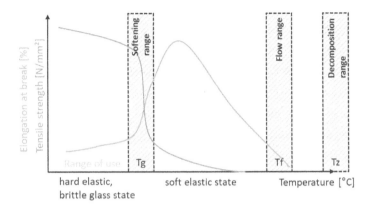

Figure 3.12: Operating temperature range of amorphous thermoplastics (adapted according to [1]).

the softening range (Figure 3.12) and thus below the glass transition temperature. In this range of use, the amorphous thermoplastics have a high tensile strength with quite low elongation at break.

In semi-crystalline materials, the amorphous phase only makes a minor contribution to the structural strength. The mechanical behaviour is much more influenced by the crystalline phase. Below the glass transition or softening range, there are high tensile strengths with low elongations at break. The semi-crystalline thermoplastic is particularly brittle there. A difference to the amorphous thermoplastics occurs when the softening range is exceeded, as the crystalline areas present in the polymer only melt in the crystallite melting range (Figure 3.13). The still existing crystal structures stabilise the polymer and ensure continued high strength. At the same time, the already softened amorphous sections in the polymer ensure a desired ductility of the material. Accordingly, the tensile strength drops only slightly, whereas the elongation at break goes up noticeably. This is the typical range of use for semi-crystalline thermoplastics. However, some plastics such as PET, PBT and PA are also used below the glass transition temperature. If the melting temperature of the crystals is exceeded, the strength of semi-crystalline thermoplastics also drops and the elongation at break rises sharply. The flow range of the material is reached almost immediately and easy processing is possible. Here, too, the material must not be heated too high in order to avoid decomposition processes.

The temperature-dependent range of use does not depend on whether the plastic is amorphous or semi-crystalline. Rather, it depends on the characteristic temperature values of the respective material. For example, amorphous plastics often have a high glass transition temperature. In the case of semi-crystalline plastics, the glass transition temperature can also be significantly below room temperature, as the melting temperature is correspondingly higher.

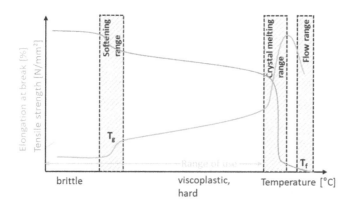

Figure 3.13: Operating temperatures of a semi-crystalline thermoplastic (adapted according to [1]).

3.6 Properties of elastomers and thermosets

Unlike thermoplastics, elastomers and thermosets cannot be melted or dissolved. This is due to the cross-links between the different polymer chains. In principle, each elastomer and thermoset is a very large molecule because all the polymer chains are linked together by cross-links. These cross-links are the result of a chemical reaction during processing. In the case of thermosets, this chemical reaction is called polymerisation. Often it is only during processing that the macromolecules are formed.

3.6.1 Elastomers

The cross-linking density of elastomers is low, which means that there are only a few cross-linking points between the polymer chains, i.e. they are cross-linked with a wide mesh. As a result, the elastomers are not soluble, but they can swell in suitable solvents by diffusing the solvent into the elastomer. The typical application temperatures of elastomers are shown in Figure 3.14.

As with amorphous thermoplastics, there is a softening range or glass transition temperature. Elastomers are used above this softening range because the polymer chains have a high degree of mobility above T_g. This mobility gets lost, below the glass transition temperature. Underneath the glass transition temperature elastomers are hard and brittle. Here, the tensile strength is very high, but the elongation at break is very low. Elastomers obtain the desired elastic properties with high strains at low tensile strength when the glass transition temperature is exceeded. There, the chains can slide off each other and expand according to the direction of the load until a cross-linking point prevents further sliding. Thus, the service range is between the glass transition temperature and the decomposition temperature.

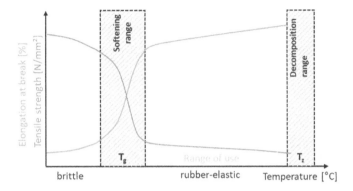

Figure 3.14: Operating temperature of elastomers (adapted according to [1]).

The behaviour shown and the explanations are greatly simplified, as elastomers never consist only of the polymer network, but additives and fillers play an essential role and influence the properties. Elastomers are used in particular in areas where high strains and rubber-elastic behaviour are required. This is the case, for example, in tubing applications, tyres and rubber bands. Elastomers are also incompressible, which is why they are well suited as sealing materials such as O-rings.

3.6.2 Thermosets

Thermosets are very closely cross-linked, i.e. they have a high cross-link density compared to elastomers. They are produced by curing, after which the thermosets can no longer be shaped by temperature or shear forces and can only be machined. The precursors are usually synthetic resins that have tri- or more functional groups, which polymerise with the help of hardeners and possibly catalysts or at elevated temperatures. These are often two- or multi-component systems. Those are mixed, shaped and then cured. Due to the close-meshed cross-linking, thermosets cannot be dissolved or swollen. Subsequent melting and forming is not possible.

As shown in Figure 3.15, thermosets can be used over the entire temperature range. After curing, thermosets are always hard and brittle. They have high strengths with a low elongation at break. Due to the close cross-linking, there is no softening or melting range. Individual chains do not exist in thermosets. It is rather a spatial close-meshed lattice of molecules.

Thermosets find applications in areas where fluctuating temperatures require a constant property profile. This is the case, for example, in the engine compartment of vehicles, in electronics applications, protective helmets, etc. Thermosets are also used as matrix material for fibre-reinforced plastic composites.

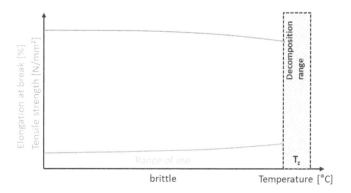

Figure 3.15: Operating temperatures of thermosets (adapted according to [1]).

3.7 Effect of the processing method on the properties of thermoplastics

The starting point for moulding thermoplastic products is usually a powder or granules. This is converted to the melt state by shear and temperature in processes such as injection moulding and extrusion. The processing methods are used to give the products their final shape. In this chapter, the possible failures due to processing will be discussed in more detail. Special processing errors can occur with the above-mentioned methods, which will also be named in the following. Depending on the material and the later component, different processing methods are available for plastics.

It is not possible within the scope of this book to provide a comprehensive overview of all processing methods for plastic products. Due to their relevance and frequency of application, the most common damage and defects in plastic products manufactured by injection moulding and extrusion are described below without claiming to be exhaustive.

3.7.1 Injection moulding

Injection moulding is a method of processing plastic granules into products. With short cycle times, large quantities of even geometrically complex parts can be produced with a high degree of reproducibility.

In the injection-moulding process, the granulate is first converted into a melt that is as homogeneous as possible by temperature and shear in a heated cylinder with a rotating screw. In the injection phase of the process, the melt is forced into the mould cavity under high shear stress by the forward motion of the screw. During the holding pressure phase, the melt is cooled by the mould temperature and the final product is ejected. Further information about injection moulding can be found in Chung [63].

3.7.2 Processing quality of injection moulding

Various defect patterns can occur during the processing of thermoplastics by injection moulding. A distinction can be made between surface and internal defects. Surface defects can often be detected with the naked eye or with low magnification. For internal defects, further microscopic or physical examination methods are required. Table 3.3 gives an overview of the most common damage cases and setting parameters that influence these damages.

A processing defect cannot always be clearly attributed to a faulty setting in the manufacturing process. Similar damage patterns can also be caused by different reasons. The quality of injection-moulded components is influenced by many parameters during processing and each parameter can cause or promote damage.

In the following, selected errors are examined in more detail and illustrated with examples.

Sink marks

On the component surface, sink marks appear in the area of material accumulations, especially with large wall thicknesses. This is caused by volume contraction during cooling. As an example Figure 3.16 shows the formation of a sink mark on a rib. A too high melt and/or mould temperature as well as a too low holding pressure or a too short holding pressure time favours the formation. As a remedy, mass accumulations can be avoided by design and the sprue position and the sprue cross-section can be optimised by design.

Record grooves effect

The surface of the component shows fine grooves perpendicular to the direction of flow, reminiscent of a record groove (Figure 3.17). This phenomenon occurs due to an excessively high cooling speed, which freezes the edge layer. This is no longer conveyed to the wall and new mass flows around it (Figure 3.18). In the holding pressure phase, the surface layer has already solidified so that the grooves cannot be closed. The remedy is a higher injection speed and a higher injection pressure as well as higher melt or mould temperature.

Table 3.3: Overview of processing errors and influencing parameters.

	Processing error	Mould temperature control	Melt temperature control	Melt plastification / homogenisation	Mould design	Holding pressure phase	Injection pressure / injection speed	Clamping force
Surface defects	Insufficient mould filling	X	X	X	X	X		
	Sink marks	X	X		X	X	X	
	Record groove effect	X	X				X	
	Warp	X	X		X	X	X	
	Diesel effect				X		X	X
	Overmoulded part	X	X		X		X	X
	Underfilling	X	X		X			
	Fibre/colour streaks	X	X	X			X	
	Gloss deviations	X	X	X				
Internal defects	Inhomogeneities		X	X				
	Voids	X	X		X	X		
	Orientations, stresses	X	X	X		X		
	Stratification	X	X					
	Cold weld lines	X	X	X	X		X	
	Impurities		X					
	Bubbles		X	X	X			X
	Free jet formation		X		X			

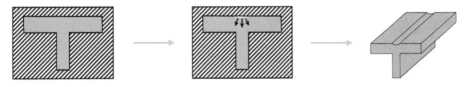

Figure 3.16: Formation of a sink mark.

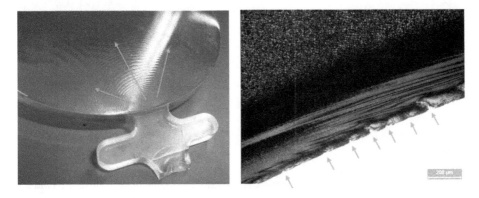

Figure 3.17: Record groove effect on a PC component (left) and thin-section images of a PVDF component (right).

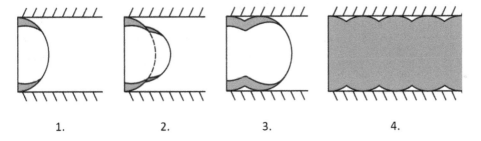

1.   2.   3.   4.

Figure 3.18: Origin of the record groove effect.

Inhomogeneities in the material

Various defect patterns can be summarised under inhomogeneities in the material. In particular, these include combustion and colour streaks, granulate that was not melted, as well as fibre and filler streaks. Some examples are shown in Figure 3.19. This is often due to an insufficiently plasticised material in the screw. This can be caused, for example, by an unfavourable screw design for the material used or incorrect parameters of a heating zone. Examples of this type of damage can be found in Sections 5.3 and 5.11.

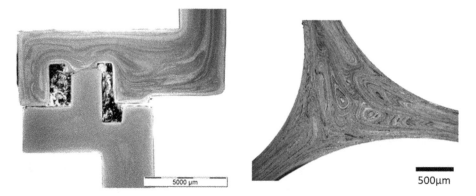

Figure 3.19: Inhomogeneities in a welded PE tank (left) and PA electronic connector in polarised transmitted light (right).

Voids

The inside of the component has cavities, so-called vacuoles or voids. These are characterised by a fissured surface. A vacuum prevails inside the part during its formation. Due to the shrinkage of the thermoplastic, the material is torn apart at greater wall thicknesses, which creates the voids. Void formation is strongly related to the shrinkage potential of the material and is therefore more pronounced in semi-crystalline thermoplastics than in amorphous thermoplastics. Possible remedies are to increase the holding pressure or holding pressure time, to reduce the difference between melt and mould temperature, to inject the component in areas with greater wall thicknesses and to set a larger melt cushion.

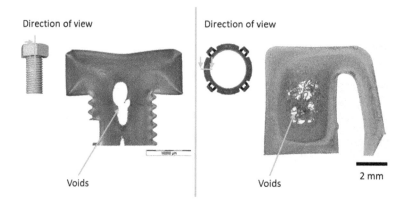

Figure 3.20: Void formation within a screw head made of PP and a component made of PA6-GF30.

Figure 3.20 shows light microscope images of components with voids. In contrast to voids, gas bubbles, for example, have a smooth surface and are formed by trapped air, moisture or low-boiling substances. Examples of this type of damage can be found in Sections 5.1, 5.7, 5.14, 5.15, 5.18 and 5.21.

Orientations and inner stresses

Rapid cooling and high shear rates of the melt causes stresses to form within the component, which can lead to cracks. These are often caused, for example, by media contact and are particularly visible in transparent materials. Internal stresses can be depicted in light microscopic thin-section images in polarised transmitted light, shown in a PC component in Figure 3.21.

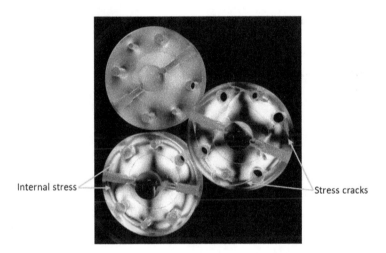

Internal stress

Stress cracks

Figure 3.21: PC component in polarised transmitted light with visible internal stresses.

The internal stresses can be prevented by a slower cooling of the melt using a higher mass and mould temperature. The position of the sprue can also influence the formation of internal stresses. Examples of this error type can be found in Sections 5.9, 5.13 and 5.15.

Cold weld lines

When two melt fronts flow together, weld lines can form. This occurs in particular when the melt fronts have already cooled down too much to join homogeneously at the confluence point (Figure 3.22). On this area, additional sink marks or notches can

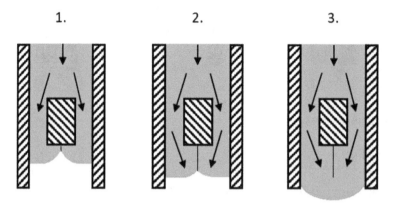

Figure 3.22: Flow path of the melt front after an obstacle.

form which are visible from the outside. The mechanical strength is greatly reduced in this area.

This can be remedied by increasing the melt or mould temperature and increasing the injection speed and holding pressure. The position of the sprue can also be changed by design so that the melt fronts meet in non-critical areas.

Free jet formation

During the injection-moulding process, no displacement flow forms in the mould and the strand of material flows uncontrolled into the cavity (Figure 3.23). This results in serpentine, rough surfaces and a layer formation on the inside (Figure 3.24).

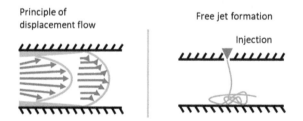

Figure 3.23: Displacement flow compared to the free jet.

The cause is usually the design, position and construction of the sprue and mould cavity. A sprue that is too small or sharp edges at the transition from the sprue to the part can also cause a free jet. A melt or mould temperature that is too low and/or an injection speed that is too high can favour this behaviour. At a glance, the examples illustrate that one incorrectly set parameter can lead to several different error pat-

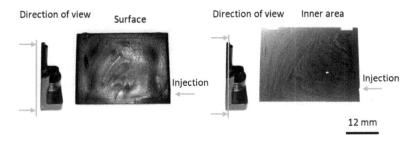

Figure 3.24: Serpentine lines on the surface (left) and stratification inside (right).

terns. A processing error can also be caused by several incorrectly set parameters or combinations of these. Further information about Troubleshooting in injection moulding can be found in Kerkstra and Brammer [23, 62].

For further literature on this chapter see [13, 16, 22].

3.7.3 Extrusion

In extrusion the plastic is melted and continuously fed through a cylinder using a rotating screw. An extruder can be used to implement various processing steps in plastics processing. They can be used for mixing, compounding, homogenising, degassing and granulating. Finished products such as profiles, pipes, seals and films are manufactured with the help of extruders. Extruded products are characterised by the continuous production line and, compared to injection moulding, are not individual parts. Portioning into an appropriate manageable size is carried out after the extrusion process.

In addition to direct product manufacturing through extrusion, for example, extrusion blow moulding can be used to realise other products through a connected discontinuous work step. In this process, a preform is continuously produced by the extruder, which is then blown into a mould by means of blowing air in a discontinuous, cyclical step. This allows hollow bodies to be produced for packaging, transport and storage articles as well as bottles, canisters or barrels. Further information about the extrusion process can be found in Chung [64] and Rauwendaal [65].

3.7.4 Processing quality of extrusion

During the extrusion of plastics, various defect patterns can occur that reduce the quality of the product. Due to the number of different extrusion processes and the possibilities of varying the process parameters in the different processes, it is often difficult to find the right reason for the defect pattern in the extrusion process, sharp defect separation is often difficult due to the continuous process, as the sharp effect on the products is missing. In the following, various error patterns will be shown as examples and possible causes and remedial measures will be given. Similarly, to the injection-moulding process, the scope of this book makes it impossible to go into detail on every possible fault, its cause and remedy of the extrusion process. A detailed description of possible error patterns at extruded products can be found in Del Pilar Noriega and Rauwendaal [9].

Polymer degradation

A typical defect of extruded plastics is an excessive degradation of the polymer. This can be determined analytically by determining the molecular weight or excessively high MFR values of the extruded plastic. The degradation of the polymer chain can take place by thermal, chemical and/or mechanical degradation, but often it is a superposition of all processes. Thermal degradation of the polymer chain is particularly due to excessively high temperatures during the extrusion process. Excessive mechanical stress on the polymer chains due to high shear or tensile forces can lead to a break in the polymer chain backbone and thus to mechanical chain degradation. Chemical chain degradation refers to a reaction of the chain with other chemicals that lead to a shortening of the polymer chain. These can be, for example, additives in the plastic or lubricants. In particular, high humidity can lead to chain degradation in polycondensates. In order to reduce chain degradation, the residence time in the extruder should be shortened and the distribution of residence time should be limited. Reducing the processing temperatures and especially the temperature peaks as well as eliminating substances that promote a degradation process also reduces the degradation behaviour.

Voids, gas inclusions and air entrapment

In the case of cavities or gas inclusions, it is first important to find out what the actual defect is. Voids and gas inclusions can be caused by moisture, solvents, volatile chemicals, air inclusions or degradation. Excessive moisture can be reduced by improving the drying of the granulate. Solvents and volatile chemicals can be introduced into the process by contamination in the extruder as well as by the polymer itself.

Air entrapment in the extrudate is created by air that is drawn into the extruder by the input material. The compression in the extruder can cause the strand to burst open when the extrudate is discharged due to decompression. If this does not happen, there are air bubbles in the product, which then also often has to be discarded. Normally, the air is pushed out of the solid bed in the feed area by the compression. If this does not succeed, various possibilities can remedy the situation. Better compaction of the material can be achieved, for example, by changing the temperature control or by increasing the pressure at the extruder outlet. It is also possible to change the particle size and shape of the input material or to prevent air inclusions with the help of a degassing zone.

Examples of this type of damage can be found in Sections 5.8 and 5.10.

Specks and impurities

Impurities and specks are problems that occur especially in very thin extrudates, films and transparent materials. In particular, such impurities are a common sight when processing recyclates. The specks can be of any colour and can often be easily distinguished from the plastic material (Figure 3.25).

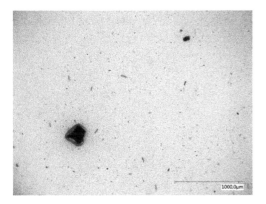

Figure 3.25: Specks in a transparent recycled PE-LD film.

The specks are caused by contamination of the starting material or foreign particles, high temperatures and temperature peaks (gel-formation) as well as stagnation in the extruder and thermal instabilities of the plastic. An example of this damage pattern can be found in Section 5.17.

Discolouration

Discolouration or uneven colour distribution deteriorates the optical properties of extruded products. Discolourisation can be caused by mixing problems in the extruder, a variation in the colour masterbatch or problems with the compatibility between colour masterbatch and polymer.

Lines in extruded products

Lines can be caused by processes in the extruder or by downstream equipment. In particular, lines in the direction of extrusion on the surface of a product are often caused by defects in the exit section of the extruder or by downstream equipment such as a cooling section.

A special form of lines are weld lines. These are located inside the extrusion strand and can be seen, for example, in cross-sectional images (Figure 3.26). In some cases, the weld lines extend over the entire cross-section, which means that a line can also be seen on the outside of the strand. The weld lines are formed by bars in the extruder, especially for hollow or multi-layer products, where the melt is separated and recombined after the bar. The polymer chains need a certain amount of time to form entanglements again and to achieve a completely homogeneous melt. The duration of this "healing process" increases as the viscosity of the melt increases. To reduce the formation of welds, the dwell time after a bar can be increased or the duration of the "healing process" can be reduced. This can be done, for example, by increasing the temperature to reduce the viscosity and accelerate the "healing process".

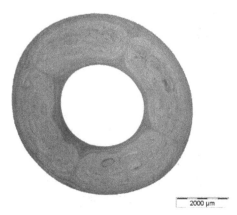

2000 µm

Figure 3.26: Illustration of weld lines.

Sharkskin

Sharkskin is manifested by a uniformly ribbed surface deformation perpendicular to the extrusion direction. Sharkskin is formed in the die area or at the exit and is dependent on temperature and extrusion speed. It is assumed that the formation of the sharkskin is caused by the rapid acceleration of the surface layers of the extrudate as the polymer leaves the die. If the stretching speed is too high, the skin can fail and form the typical grooves (Figure 3.27). Sharkskin can be avoided by increasing the die temperature and reducing the extrusion speed. Also, the use of an external lubricant can reduce the formation.

Figure 3.27: Sharkskin on the surface.

Melt fracture

A melt fracture does not refer to defects on the surface of the extrudate but involves the entire extrudate body. Various examples of melt fracture are shown in Figure 3.28. Several mechanisms have been designed for melt fracture that describes the failure pattern. Most likely, the melt fracture is caused by critical elastic deformations in the input zone, critical elastic strains as well as slip-stick effects in the nozzle. The failure pattern can be reduced by streamlining the nozzle, increasing the nozzle temperature, reducing the molecular weight or viscosity of the polymer, increasing the cross-sectional area of the outlet or using an external lubricant.

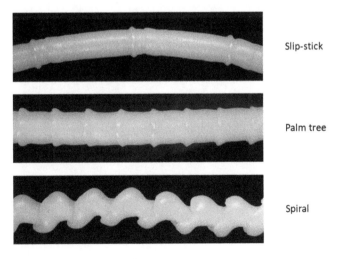

Slip-stick

Palm tree

Spiral

Figure 3.28: Different melt fracture patterns.

3.7.5 Further processing methods

Further processing of plastics can also cause defects in the product. In the welding process, for example, if the energy input is too low, the weld will not be sufficiently good and the desired interlocking of the polymer chains of the joining partners will not take place. The possible or required strength of the welded joint will not be achieved, which may lead to failure of the product in later use. If the energy introduced into the welding process is too high, the molecular chains may break down. This can lead to gas bubbles or other inclusions in the area of the joint. Mechanical machining of plastic products, such as milling and sawing, can also generate excessive heat in the plastic, resulting in localised defects. When bonding plastic parts, it is important to ensure that the parts to be joined and the adhesive are compatible. In particular, the surface of the parts must be suitable for bonding. This shows that the previous processing steps can also be decisive for the further processing of the plastics. It is not always possible to determine which processing step caused the damage. In addition to processing, defects introduced by the material or filler, or a combination of material, filler, processing and further processing, can also be the cause of damage. In the course of the life cycle, further aggravating influences are added, which can also lead to a failure. Failure analysis of plastic products is therefore difficult and complex and requires a great deal of experience and knowledge in various areas of the life cycle of plastics.

Rainer Dahlmann, Edge Fischer, Jan Buir, Christoph Zekorn,
Sabine Standfuß-Holthausen, Meike Robisch, Michèle Marson-Pahle,
Christiane Wintgens, Hakan Çelik

4 Overview of most important analytical methods

This chapter presents established methods for carrying out analyses as part of the damaged analysis. This is only an excerpt of possible analysis methods.

4.1 Microscopic analysing methods

4.1.1 Light microscopy

Significance of failure analysis

One of the most important tools for carrying out failure and damage analysis is the technique of light microscopy and its associated preparation methods. At the beginning of a damage analysis is the comprehensive documentation of the damage pattern itself. The damage image is usually observed with the naked eye, a magnifying glass, a macroscope or microscope. In addition to the initial assessment of the damage, light microscopy helps to identify and rate typical processing faults and defects, which can be directly linked to the cause of component failure. The analysis not only provides information about specific external characteristics of investigated objects but, in particular, also insights into the internal characteristics of plastic products. These are usually influenced by the material composition and its processing. In addition, use-related influences or stresses can also lead to a significant change in internal (structure) and external (topography) characteristics. Here, too, light microscopy is a valuable tool, for example, to reveal previous medial interactions or ageing effects in the material.

Using light microscopic analysis, the following conspicuous features and defects, among others, can be visualised (see Table 4.1). The analysis of stresses/orientations as well as of shear and edge zones is typically carried out in transmitted light with the use of polarising filters. Here inhomogeneities can not only be shown in the transmitted light method but also in the reflected light method.

Typical conspicuous features such as cavities and blowholes, cracks, white fracture, inadequate form filling, weld lines, shrinkage and filler distribution are examined with the aid of reflected-light microscopy. Small differences in height, such as

https://doi.org/10.1515/9783110785647-004

those directly at weld lines, can be visualised by using differential interference contrast microscopy (DIC).

The sample preparation must be carried out in correspondence with the analysis methods chosen here. This is explained in more detail in the following text.

Table 4.1: Typical defects and conspicuous features.

Defects and conspicuous features	Analysis method
Stresses/orientations	transmitted light method, use of polarising filters
Inhomogeneity	transmitted light method
Cavities/blowholes	reflected and transmitted light method
Cracks	reflected and transmitted light method
Shear and edge zones	transmitted light method, use of polarising filters
Stress whitening	reflected light method
Inadequate form filling	reflected light method
Weld lines	reflected light method, differential interference contrast microscopy (DIC)
Shrinkage (e.g. spot)	reflected light method
Filler distribution	reflected light method

Functional principle

In light microscopy, images are generated by using light-refracting glass lenses, the combination of which produces a magnifying effect on the sample image. The working principle corresponds to Figure 4.1.

In reflected light microscopy, light passes through the various apertures and the collecting lens and is focussed on the object reflected from the sample surface and then enters the objective. In transmitted light microscopy, the light shines through the sample and then enters the objective (thin section examination). The light path includes a converging lens that focuses the light beam coming from the light source. The field diaphragm (iris diaphragm) limits the beam diameter and eliminates scattered light, while the aperture diaphragm helps to create contrast. The condenser bundles and focuses the light on the sample plane. After the transmission through the sample, the light enters the objective, which produces a side-inverted magnified image of the sample, and then the ocular.

Reflected light microscopy is carried out on authentic part surfaces or cross-sections and can be used in particular to analyse the distribution of fillers and reinforcing materials and to investigate the amount of cavities/blowholes, cracks and inhomogeneities. In plastics analysis, transmitted light microscopy is usually carried out on thin-film preparations that can be transmitted by light [49, 51].

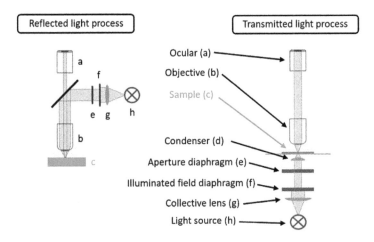

Figure 4.1: Schematic function of the light microscope.

The use of polarisation filters in transmitted light microscopy is a very often used technique to assess internal stresses and orientations in plastic products (see Figure 4.2). The visualisation of these properties is based on the principle of birefringence. In this application, rotatable polarising filters are inserted into the beam path between the light source and sample (polariser) and the sample and occular (analyser). The polariser generates linearly polarised light with just one direction of oscillation. Because of the birefringence properties of many plastics, the linearly polarised light in the sample is split into two partial beams. Depending on the local anisotropy within the sample, the partial beams experience a different propagation speed and thus a phase shift. Interference of the phase-shifted partial beams then occurs at the analyser (90° rotated), resulting in the characteristic isochromats. This is why inner stresses and orientations show up in the form of varying interference colours whose frequency or density of colour changes is usually used for qualitative assessment purposes. In this way, the spherulitic structures in semi-crystalline materials can also be examined [48].

In addition to "normal" bright field illumination, dark field microscopy can also be useful in plastic damage analysis. Here, the light radiating from the sample towards the objective is first guided past the objective opening in the form of a hollow cone of light, so that the object appears dark to the observer. However, if there are fine pores, fillers or reinforcing materials in the material, these scatter the light locally in the direction of the lens opening, resulting in a corresponding image contrast. The use of these methods should be chosen depending on the features to be depicted. As it is easy to set up, bright-field microscopy is the most used standard to analyse inclusions, pigment distributions or voids in fibre-reinforced plastics. In the case of filler, crack or pigment analysis, dark-field microscopy can have additional benefits.

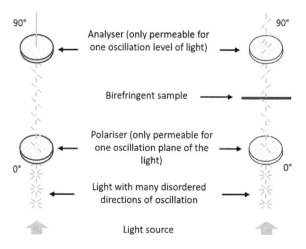

Figure 4.2: The principal of polarisation microscopy.

Another method in reflected light microscopy is DIC, which is used, for example, to display small differences in height or to display crystalline structures in relief. With this method, differences in the optical path length at the observed object surface can be converted into colour differences in the image through light interference. A characteristic relief effect is created, which gives a very plastic impression of the topography of authentic moulded part surfaces or even polished cross-section surfaces. In the qualitative evaluation of injection moulded parts, for example, this makes it very easy to highlight shear zones, fibre orientations or weld lines in cross-section [1, 7, 46, 47, 48, 49, 50, 51, 53].

Requirements for sample quality and preparation

Fracture analysis is a method that usually has the fracture opening as the only preparation step. The results can provide information about the internal and external properties of the plastic components.

To analyse the internal properties, such as the component morphology, plastic products usually have to be prepared for light microscopy. This is only possible with suitable cross-sectional samples. There are various preparation possibilities, which are explained in more detail in this chapter. For analysis using the reflected light method, cross-sections are produced. Here, sample segments are embedded in a cold-curing epoxy resin and ground and polished after a curing period so that they can then be examined microscopically. For the transmitted light method, there are two different preparation techniques for producing light-optical transmissive samples. On the one hand, solid polished cross-sections are first glued on a glass carrier. After curing, the block is sawn off close to the glass carrier and the sample is ground down thinly towards the glass plate in analogy to the cross-section preparation until it can be transmitted by light. The target thickness can be between a few hundred μm (thick section) and a few 10 μm, depending on the material, colourants and fillers, as well as on the task to be answered.

On the other hand, thin sectioning can be realised using microtomy. To do this, the specimens are moved past a blade in a repetitive motion and moved closer to the blade by the amount of the desired cut thickness in each cycle. This results in thin sample chips that are transferred to a glass slide for transmitted light microscopy and enclosed with a drop of immersion oil and a cover glass. It is not always advisable to cut materials with a high filler content or fibre-reinforced plastics using microtomy. The fillers and reinforcing materials can be pulled out of the plastic matrix during cutting, so the thin cut does not necessarily represent the actual state of the sample. These filled and reinforced materials would also cause the knives used to wear out prematurely, so they would have to be changed more frequently. The tools used play a major role, such as the type of knife that is used to produce the thin sections. Depending on the material stiffness and component thickness, different materials such as stainless steel, glass and diamond can be used for the blade. For detailed applications in practice with the named preparation methods, further literature is recommended [47, 51, 53].

Comparing the preparation methods of grinding and cutting, it can be stated that thin cutting with microtomy offers a significantly faster way of assessing the properties of the plastic. Depending on the material and filler, however, it is not always suitable. A sample geometry that is too complex also makes it difficult to create the cuts. For less elastic or reinforced plastics as well as for complex specimen geometries, preparation by grinding is recommended. This method offers the advantage that it is low in deformation and, compared to the sectioning technique, relatively large specimen areas can be examined.

Typical results of failure analysis

Typical results of light microscopic analysis are shown in the next two composed figures. Since this method provides a qualitative result, reference samples, experience as well as knowledge of the methods used in processing play a significant role in the evaluation of the results. Figure 4.3 shows typical results of transmitted light microscopy (partly recorded in polarised transmitted light). In addition to an extruded pipe (thin cut, left side), which shows inhomogeneities in the form of poor mixing of the colour masterbatches, a sample with an inclusion (thin section, recorded in polarised light), as well as a sample with clearly visible stresses/orientations are also shown (thin section, recorded in polarised light). The presence of inhomogeneities in the form of layers and especially inclusions can significantly reduce component stability and can act as a weak point. The visible stresses/orientations can also be generated and intensified by various influences during processing or component use. For example, strong molecular orientations can form and solidify, leading to further stresses. This can affect the material strength and under certain conditions, such as media contact, lead to the formation of environmental stress cracking (ESC).

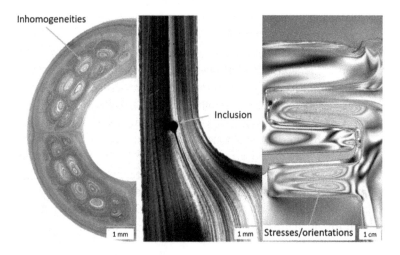

Figure 4.3: Typical results of transmitted light microscopy (left: pipe, mid: inclusion, right: stresses/orientations).

Figure 4.4, on the other hand, shows typical images from reflected light microscopy. On the one hand, a section of a glass-fibre filled sample with cracks is shown (cross-section, left side). This sample shows a failure of the component, which had a non-optimised component design and failed an internal test. On the other hand, an analysis of a fracture examination is shown. The image of the fracture surface shows that there is an inclusion at the origin of the fracture, which was the cause of the component failure (non-prepared sample). The image on the right-hand side shows shrinkage cavities inside a fibre-reinforced component. In this case, the too low-temperature control during the process (melt and mould) played a major role in the formation of the flaws, which can also act as a weak point, depending on the position of the component and the extent of the defect (cross-section).

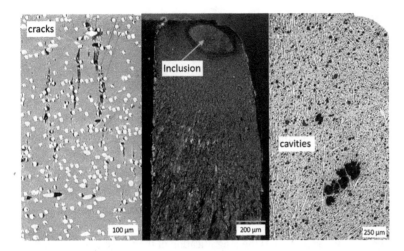

Figure 4.4: Typical results of reflected light microscopy (left: cracks in an FRP, mid: fracture analysis, right: cavities due to shrinkage).

Further examples can be found in chapters 5.3, 5.8, 5.12, 5.15 and 5.19.

For further literature on this chapter see [15].

4.1.2 Scanning electron microscope (SEM)

Significance for failure analysis

Among the microscopic methods, electron microscopy is characterised by a wide range of magnifications and at the same time high lateral resolution down to a few nanometres. The technique benefits from a very high depth of field, which is approx. 100 times higher compared to conventional light microscopy [50]. Furthermore, the imaging is not influenced by the inherent colour or the reflection behaviour of objects under examination. This combination of features makes SEM ideal for high-resolution imaging in the context of damage analyses on plastic components. For example, the examination of component surfaces can provide information about manufacturing-related characteristics, chemical attack or age-related damage. Furthermore, electron microscopy allows the evaluation of fractured surfaces and identification and characterisation of material inclusions or fillers that may be directly related to component damage. A fracture surface often allows conclusions about the failure origin and progression and enables the cause of the damage to be defined.

Another helpful SEM feature in the context of failure analysis is that differences in material density can also be visualised and analysed. This means fillers or reinforcing materials as well as chemical stained polymer phases. A quantification of such phases with regard to their shape, size and distribution is often needed to conclude about the component quality.

Functional principle

SEM uses a focussed electron beam to generate images of a sample in a vacuum chamber. Electrons are first generated at a cathode and accelerated towards the sample by an applied accelerating voltage. The entire system works under vacuum conditions to avoid scattering interaction between the electron beam and gas atoms/molecules. On their way from the cathode through the column down to the chamber, the electrons are formed and focussed to a small beam using electromagnetic lenses. The focussed beam is guided over the surface to be examined in a line-like manner via a scanning unit. The size of the sample area scanned by the beam defines the magnification of the image on the monitor.

Various interactions occur on the surface of the sample when it comes in contact with the primary electron beam, some of which can be used to generate images (Figure 4.5). Depending on the primary energy of the electrons and the density of the material, the measurable interaction is limited to an excitation volume with a depth of 2–20 μm [39]. On the one hand, electrons are knocked out of the sample atoms (secondary electrons/SE) and, on the other hand, there are electrons of the primary electron beam that are backscattered elastically by the atomic nuclei of the sample material (backscattered electrons/BSE). In the course of the formation of SE, space changes take place in the affected atomic shells, resulting in characteristic X-ray quanta that can be identified with suitable spectrometers in terms of their energy (EDX, see Section 4.2.2) or wavelength (WDX) and assigned to specific elements.

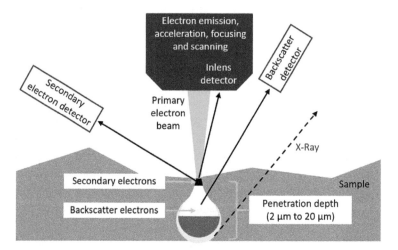

Figure 4.5: Properties of an electron beam hitting matter [40].

Since the SE are generated close to the sample surface, SE images are well suited for recording the topography of a sample surface. The contrast is created by shadowing effects. SE that come from a valley or an area facing away from the detector reach the SE detector at a smaller intensity than those that come from a mountain or an area facing the detector. In this way, mountains or areas facing the detector appear brighter.

BSE are elastically backscattered from the atomic nucleus. This makes them usable to generate a material contrast depending on the atomic nucleus size. Here, the statistical probability of a collision with a primary electron increases with increasing atomic nucleus size. Since a larger number of atoms reflected from a region means a higher intensity, regions consisting of atoms with larger nuclei appear brighter than regions consisting of atoms with smaller nuclei. This method is therefore a suitable means of identifying and visualising fillers and reinforcing materials in plastics.

Requirement for sample quality and preparation

Electron microscopy has certain requirements for the specimen. In general, the solid sample or at least its surface must be electrically conductive. The electrical conductivity is important, as absorbed electrons charge an electrically isolated sample and then scatter further electrons from the beam. In addition, the specimen must be vacuum stable to avoid volatile components to outgas from the material. In the context of damage analyses, this requirement can be a critical point depending on the sample history (e.g. sealing ring with oil contact).

For the reasons mentioned, plastic samples that are normally not electrically conductive are vapour-deposited with a conductive material (gold, carbon or platinum) by means of a sputtering process. The vapour deposition can be done in different ways. Usually, small vacuum chambers are used in which the sample is placed and gold particles are sputtered out of the target material by igniting an argon plasma. The gold atoms deposit on the sample surface and form a conductive layer with increasing coating time. The surface topography of the sample can have a great influence on the achievable electrical conductivity. Undercuts or cracks on the sample surface bear the risk of a locally interrupted conductive coating and lead to impairments in the image due to local electrostatic charging effects. Experience shows that in such cases several coatings at different angles of inclination of the sample are helpful. Furthermore, conduction of the electrical charge from the conductive surface to the specimen holder of the electron microscope must be ensured, which is usually achieved by gluing on a conductive carbon or copper strip as a conductive bridge (electrical ground) [51].

For very small samples or even dust/powders, conductive adhesive pads are used on which these samples are applied. It is important to ensure that the powder-like samples do not detach from the pad during the examination to prevent contamination of the chamber or column.

A special case of preparation occurs within the analysis of chemical-stained polymer phases. For example, elastomer phases (such as poly-butadiene in ABS) that are to be evaluated in a thermoplastic matrix can be stained with heavy metals like osmium tetroxide. Here a chemical reaction takes place in the area of carbon double bonds that causes a phase selective enrichment of Os-atoms in the rubber phase. Selectively contrasted areas increase the BSE contrast in the example to visually distinguish the elastomer phase (lighter due to Os-atoms) from the thermoplastic matrix (darker). This elaborate preparation method requires particularly smooth preparation surfaces, which can only be produced with the help of diamond microtome cuts.

Typical results of failure analyses

Using SEM, images can be generated either surface-sensitive or material-sensitive. Two images of the same defect can be seen in Figure 4.6. In the BSE image, the glass fibres in the inclusion appear much brighter than the surrounding material. This is because the chemical elements of the glass fibre (e.g. Si, Al, O) have larger atomic nuclei than the surrounding carbon of the polymer matrix. The SE detector image clearly emphasises the topographical properties. Here there is a shadow effect due to the elevations facing the detector.

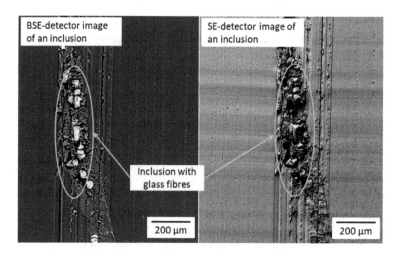

Figure 4.6: Comparison image of an inclusion in PC, with the BSE (left) and SE (right) detector.

Fracture surface analysis is one of the typical examples that can be performed using an electron microscope. The fracture surface analysis is used to narrow down the conditions at the moment of failure. To gain a brief insight into the subject, an example is shown in Figure 4.7 on the left-hand side. The image on the left-hand side shows a section of a fracture surface in the transition to the so-called residual fracture on a polycarbonate sample. Here are rest lines that merge into the rest fracture area. The rest lines may or may not be an indication of an alternating load that has led to the fracture. The transition between the rest lines and the residual fracture area with the tips represents the moment when the component was weakened to such an extent that it broke through in a final continuation of the crack. The tip lines provide further information about the direction of crack propagation, as the point on the direction of fracture propagation is in the shape of an arrow.

Another typical example is the analysis of fillers or phases that are too small for light microscopy. On the right-hand side of Figure 4.7, the butadiene phase of an ABS material is shown. The sample was prepared using chemical staining as described above. On the cut surface, the denser areas of the polybutadiene phase stained with osmium can be well visualised with BSE detector. The images are usually inverted for a subsequent grey value correlation, which is why the denser areas appear dark and not bright. With the help of the grey value correlation, the average butadiene phase diameter can be determined on the basis of several image data values, which have an influence on the toughness of ABS.

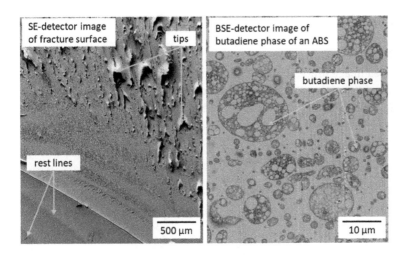

Figure 4.7: Fracture surface of a vibration-loaded PC sample (left) and butadiene phases of an ABS (right).

Further examples can be found in chapters 5.10, 5.17, 5.18, 5.23 and 5.24.

4.1.3 Industrial computed tomography (CT)

Significance for failure analysis

Conventional microscopic imaging techniques, such as light or electron microscopy, allow two-dimensional structural investigations on sample surfaces or cross-sectional areas, or on the basis of thin-film preparations that can be transmitted. However, statements on three-dimensional structures or on the spatial orientation of features of interest (phases, fillers, pores, defects) cannot usually be made by these means. This is where industrial X-ray CT can show its benefits, by generating three-dimensional, quantifiable volume information.

X-ray CT has a long history of application and has become an indispensable tool, especially in the field of human medicine, which is certainly also related to the rapid advances in the field of information technology and data processing. However, the technique plays an important role in failure and damage analysis of plastic products as well. Because of the possibility of basically non-destructive and yet material-penetrating imaging, especially unique specimens can be subjected to a thorough defect analysis (e.g. course of crack formations or leakages, localisation of cavities or material inclusions) without permanently changing the object under examination. CT is therefore often a very suitable technique to identify or define particular regions of interest (ROI) on and in an object of interest for subsequent precision analyses (e.g. spectroscopy).

Functional principle

In summary, the process initially involves radiographic X-ray analysis, followed by software-supported image processing for the reconstruction of two-dimensional layer images, which are in turn used for three-dimensional viewing and evaluation of the scanned sample volume. Figure 4.8 illustrates the three essential hardware components of a computer tomograph.

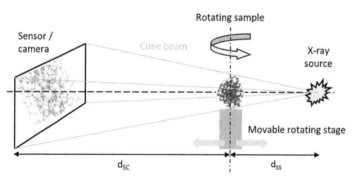

Decreasing d_{sc}: + larger sample size possible / - scan resolution
Decreasing d_{ss}: - smaller samples needed / + scan resolution

Figure 4.8: Principal of CT-image generation.

Coming from the point-shaped X-ray source, a cone-shaped X-ray beam passes through the specimen fixed on the rotating stage. The X-ray radiation is attenuated to varying degrees. The strength of the attenuation is determined, among other things, by the thickness of the specimen (transmission length), the local density, respectively, the atomic number of the material or phases through which the radiation passes. Behind the specimen there is an X-ray sensitive detector that converts the incoming X-ray intensity into a digital X-ray projection image. When performing a CT measurement, usually at least a half rotation (better a full rotation) of the sample under examination is performed in the beam cone and a finite number of X-ray projections are acquired in defined step widths of the rotating stage (e.g. 0.3°). For a 180° scan and a step width of 0.3°, 600 X-ray projections are available for further processing. Mathematical algorithms are used to convert the X-ray projection images into 2D reconstruction images or, in the simultaneous processing of all reconstruction images, to 3D volumes. The grey values assigned to the image points essentially correspond to the local density or atomic number of the corresponding material within the sample [54].

Requirement for sample quality and preparation

According to the current state of the art, the size of the examination volume and the achievable detail resolution (μm/voxel) are inversely proportional to each other, which relativises the reputation of the method as a "non-destructive examination" depending on the task. This circumstance is due to the fact that CT measurement requires a complete image of the sample on the detection surface of the camera. At the same time, to increase the imaging size on the camera (=magnification), the distance between the sample and the X-ray source must be reduced. The largest dimension of the examined sample thus indirectly determines the achievable resolution. Thus, a high scan resolution in the lower single digit μm range, as required for visualising individual glass fibres in the component structure, for the reasons mentioned above, must be preceded by a confectioning of the sample volume to an edge length of a few millimetres. With modern micro-focus or nano-focus tubes and correspondingly high-resolution detectors, resolutions of well below 1 μm can nowadays be achieved even with laboratory equipment and a correspondingly small sample. Improvements in terms of small examination volumes are promised in the future by further developments of CT systems that combine the currently established process technology and the use of a scintillator with associated lens systems to achieve optical post-magnification of the X-ray projections.

With the numerous possibilities that CT offers for plastics analysis, however, there are also material-specific application limitations here. Challenges often arise with material combinations or multiphase systems when the components contained are to be imaged or evaluated in isolation from one another. Since the imaging, contrast is directly related to the density and thus transmissivity of the material under investigation, there is a requirement for a minimum density difference. Based on experience, for polymer analytics those differences should be in the region of approx. 1 g/cm^3, to achieve a good phase or material separation. Pores in compact material (e.g. polyamide 6 with density $\rho \cong 1.14$ g/cm^3) understandably represent an ideal starting situation in this context, while, for example, short fibres of carbon in a polyethylene terephthalate matrix remain nearly invisible (CF $\rho \cong$ 1.55 g/cm^3 vs. PET $\rho \cong$ 1.38 g/cm^3).

Conversely, problems can also arise from excessive density differences, due to specific X-ray energies required for material or phase transmission. To illustrate this, consider the example of an electronic connector whose thermoplastic body contains metallic pins for electrical current flow. The X-ray voltage required for optimum penetration of the pins, or the plastic varies by a few 10 kV, which means that, depending on the scan setup, the plastic is displayed too brightly or becomes invisible due to overexposure at high X-ray voltage. In the case of the other extreme, the pins are penetrated insufficiently at low X-ray voltage (in this case appropriate for the plastic). The result can be image artefacts in the reconstruction images (e.g. locally strong over-radiation), which complicates or even prevents an evaluation of the results. However, there are increasingly viable solutions to this problem from the equipment manufacturers, for example, by combining multiple scans with different parameterisation or by combining different X-ray tubes. In addition, metallic filters (Al or Cu-Al) in the beam path can reduce the overall contrast level in the scan within certain limits, even though at the expense of imaging quality or detectability of detailed structures.

Typical results of failure analysis

In practice, plastic mouldings can be scanned and evaluated as a whole for the purposes of reverse engineering, dimensional measurements or the detection of voids, if the measuring chamber is large enough. Numerous quantitative statements can be extracted from the generated data using grey value correlation methods. Examples include the determination of pore sizes (Figure 4.9) and distributions or the determination of the orientation tensors of reinforcing fibres in plastic moulding compounds. CT results allow qualitative evaluations, such as statements on the compactness of a material or crack paths in moulded parts, which are not accessible to the user by purely external observation. Furthermore, CT is suitable for tracing assembly conditions in assemblies, such as the seating of sealing rings that are not visible from the outside.

A method useful for damage analysis for the detection of fine and therefore difficult to detect cracks, fissures, etc. uses the principle of artificial contrast enhancement to improve the detectability of phases of interest (here: air) in contrast to the plastic. For this purpose, for example, potassium iodide can be applied as aqueous solution to the presumed area of damage (e.g. visible crack on the surface) via a pipette. The solution is drawn into the finest cracks via capillary forces, causing them to become enriched with potassium iodide. Depending on the component and situation, this process can be accelerated by applying a vacuum. The potassium iodide exhibits a high imaging contrast in the CT compared to the plastic and the smallest amounts remaining in fine crevices are often sufficient for adequate detectability.

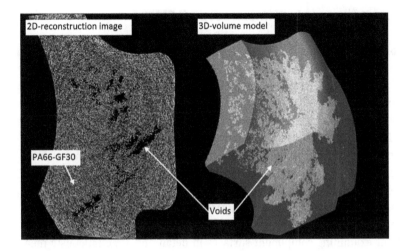

Figure 4.9: CT-analysis of injection-moulded PA66-GF30 part segment with voids.

Figure 4.10 shows an application example from industrial μ-CT, where the screwing behaviour of medical luer connections was to be investigated. The possibility of displaying entire volume or functional units sometimes enormously facilitates the traceability of given fits or assembly situations and helps in failure identification. In the example given, the three-dimensional sectional view made it possible to understand and evaluate very well the load-bearing component areas in the thread area as well as the fit of the surfaces responsible for the tightness of the system.

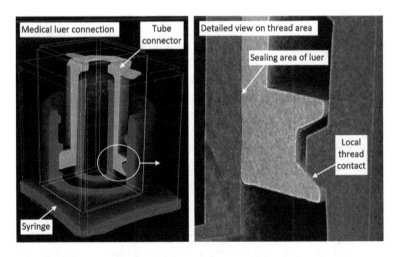

Figure 4.10: Virtual cross-section through 3D CT-volume of luer connection.

Further examples can be found in chapters 5.16 and 5.21.

4.1.4 Laser scanning confocalmicroscopy (LSM)

Significance for failure analysis

LSM combines the conventional reflective light microscopy with a successively built-up reflection image of a laser light source. The analysis allows displaying and evaluating defects or in general topological characteristics on authentic component surfaces or fracture surfaces in high resolution within the scope of defect and damage analysis, partly non-destructively. This method is therefore used to scan, measure and evaluate surfaces non-tactile. In damage analysis, the technique is used to evaluate differences in gloss, allows determination of surface qualities, analysis of moulding accuracies, layer thickness measurement on multi-layer systems with transparent layers or determination of weld lines, to name some examples.

The ability of examining component surfaces in detail and simultaneously measure them three-dimensionally is a special feature of this method, as the LSM provides exact height information for each pixel in the image. In addition to a 3D surface observation, various roughness parameters can be determined to reveal conspicuous features or differences between components (e.g. comparison between LOTs or with reference samples). Frequently, the arithmetic centre roughness value R_a, the square centre roughness value R_q and the average roughness depth R_z are determined for this purpose, as well as the peak (R_p) and valley (R_v) values. The use of various filters or a tilt or bend correction of the measured plane also offers further opportunities for evaluation. Samples can also be examined as a line scan to achieve long-distance surface information in a time-saving manner.

Functional principle

The LSM is a laser-assisted microscope that uses a laser source in addition to a light source (e.g. LED) for the analysis. The confocal microscope differs in design from a conventional light microscope with regard to the use of pinholes, which serve as a point light source or point detector. By using these pinholes, the light that does not pass the focal plane is almost completely blocked. The special feature of this method is that images are produced with comparatively high contrast. This is due to the fact that only light from a small volume around the focus reaches the detector. A simplified illustration of the confocal beam path compared to a non-confocal beam path is shown in Figure 4.11.

The analysis principle of a laser-scanning-microscope can be seen in simplified form in Figure 4.12. After passing the pinhole the light passes a semi-translucent mirror and the objective lens and reaches the sample surface, which is located in the focal plane (green arrow). The light is then reflected from the sample surface and finally reaches a photomultiplier after passing a second pinhole, which blocks the stray

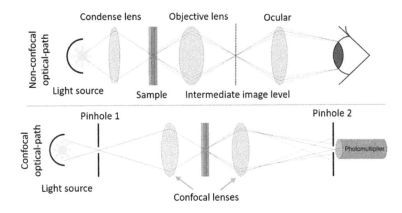

Figure 4.11: Schematic of the confocal and non-confocal beam path.

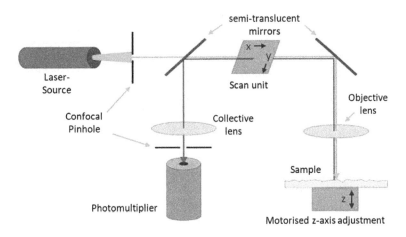

Figure 4.12: Analysing principle of the laser scanning microscope (laser beam path).

light as well as light reflected from the areas outside the focal plane. As a result, only light from a small volume around the focal point reaches the detector, producing high-contrast images in the end. To do so, the point-shaped laser beam is guided over the sample in rows and layers (variable Z-height) via the deflection unit until all surface structures have been detected and processed by the software into a cumulative image. The depth resolution is adjusted by the diameter of the aperture. The focus is also adjusted via the motorised objective lens. Based on these properties, the surface topographies and structures can be evaluated in three dimensions.

Requirement for sample quality and preparation

When analysing surfaces such as injection moulded parts, generally the sample preparation just consists of ensuring that the sample is clean and fits under the microscope in terms of its dimensions. If the area to be analysed is inside a specimen or if a crack is to be examined, of course the specimen has to be opened by means of sawing or cryo-fracturing.

The laser beam coming from the objective hits the sample surface to be examined orthogonally. For physical reasons, the imaging precision therefore decreases with increasing flank angles of structures on the surface. Consequently, undercuts cannot be imaged. When examining damaged parts, deep crack trenches can also only be imaged to the extent that light can still reflect from the ground and reach the lens. In principle, the image is independent of the inherent colour of the object. Compared to conventional light microscopy, this results in an advantage for the surface analysis of white or transparent moulded parts, for example. Extremely polished (for example metallic) surfaces with high reflection can lead to misinterpretation of the reflected laser light, depending on the lens used. This can manifest itself in supposed peaks in the roughness profile, for example [49, 51]. The limit of the working distance is 128 mm, with nanoscale resolution in the Z-plane and 0.012 μm resolution in the XY-direction.

Typical results of failure analysis

Figure 4.13 shows an analysis result of a surface of an injection-moulded component, which shows differences in gloss. The left-hand side shows the laser optical grayscale image (2D image); the right-hand side shows a 3D image. Comparatively, the 3D image shows local differences in the height of the surface. In the upper image, in the area of the gloss differences, individual positions with higher elevations (peak to valley) can be seen. Overall, the lower sample appears more even (laser image). This difference can be measured quantitatively and is non-destructive with this method. Here, the maximum height for the surface with gloss differences is approx. 30 μm, whereas the inconspicuous sample showed height differences of approx. 22 μm. An additional analysis of the roughness and the evaluation of the roughness parameters would also be optionally possible here.

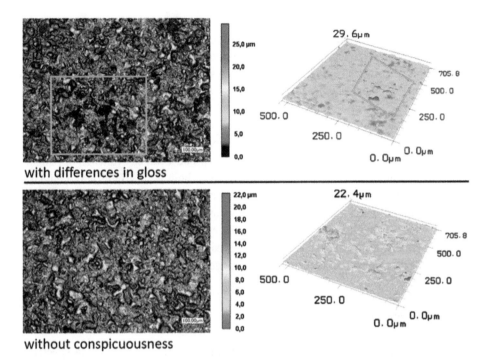

with differences in gloss

without conspicuousness

Figure 4.13: Investigation of gloss differences of an injection moulded component/comparison of areas with and without anomalies.

Further examples can be found in chapters 5.8 and 5.13.

4.2 Spectroscopic and chromatographic analysis

Defects and damage can also occur if the material itself does not fulfil the requirements of the application or the stresses in the processing procedures. Spectroscopic methods can be used to identify a large number of the ingredients in plastics, for example, by carrying out a structural analysis or an elemental analysis. To increase the information quality, chromoatographic separations are useful.

4.2.1 Fourier transform infrared spectroscopy (FTIR)

Significance for failure analysis

IR spectroscopy is one of the most important methods in plastics analysis for determining which material or polymer type is used. In this way, not only material classes or types but also, for example, material mix-ups, batch variations or impurities will be identified. The availability of extensive databases, the rapidity of the method as well as the generally low preparation effort support the high value of this method.

In this structure clarifying material identification, a variety of organic as well as semi-organic components will be recorded so that all dominant components of this type will be detected at the same time. In addition to the classic identification of materials, structural changes can also be identified by using suitable references. An example is the oxidative damage and thus ageing (triggered thermally or by UV radiation) of polyolefins in particular. In this way, the current condition of a product can be evaluated in relation to the specifically selected analysis position. This can be useful in failure and damage analysis as well as in quality assurance in the context of a material evaluation or specification. Another application example for IR spectroscopy is the time tracking of reaction processes such as the chemical conversion of diisocyanate and a diol to a polyurethane.

Furthermore, a special technique of this procedure is used to identify surface defects such as impurities due to contamination or components that have migrated to the surface.

Functional principle

This chapter describes the principle on which the analytical method of IR spectroscopy is essentially based. Furthermore, common analysis techniques and their major differentiating features are presented.

Structure clarification by means of IR spectroscopy is based on the interaction of electromagnetic radiation and molecules [41, 42], in that the radiation serves as an energy source to stimulate and thus intensify already existing oscillations in groups of molecules (see Figure 4.14).

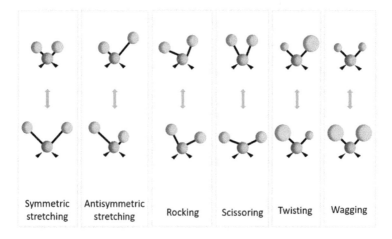

| Symmetric stretching | Antisymmetric stretching | Rocking | Scissoring | Twisting | Wagging |

Figure 4.14: Possible molecule vibration stimulation by means of IR radiation.

The essential measurand is therefore the wavelength depending on the reduction of the radiation intensity, which correlates with absorption by vibration stimulation. For plastics analysis by IR spectroscopy, two main techniques are used in practice:
- transmission method and
- attenuated total reflection (ATR).

In the transmission method, a thin sample is placed in the beam path (see Figure 4.15). Accordingly, the entire sample cross-section is analysed simultaneously. In the case of multilayer systems, therefore, the middle layers are also included, with the result being the total of all the materials in each layer. It should be noted at this point that plane-parallel specimens can cause the spectra to be inaccurate with this method, since multiple reflections result, and the resulting interference leads to errors in the detected signal.

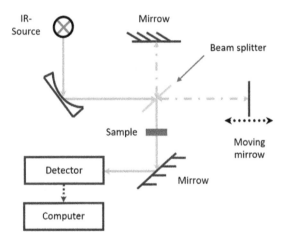

Figure 4.15: Schematic principle of operation of the FTIR analysis with the transmission measurement mode.

Depending on the equipment, gases or liquids (pure substances as well as mixtures) can also be analysed. Powdered samples must be stamped into pellets in advance with IR-inactive substances for transmission analyses. Using IR analysis in transmission, both the classical mid-infrared range (MIR; wavelengths between 2.5 and 25 μm or wavenumbers between 4,000 cm^{-1} and 400 cm^{-1}) and the near-infrared range (NIR; detection of so-called harmonics; wavelengths between 800 nm and 2.5 μm or wavenumbers between 12,500 cm^{-1} and 4,000 cm^{-1}) can be used.

The more common method of characterising plastics is the ATR method. In this technique, a measuring crystal (diamond, germanium or zinc selenide) is located in the beam path. This measuring crystal is geometrically designed in such a way that the IR radiation enters at an angle of incidence above the critical angle and is thus completely reflected at the theoretical exit surface and leaves the measuring crystal only when it hits the next interface. Immediately neighbouring the reflection surface of the measuring crystal, a decaying field is formed, which can interact with this environment close to the interface (at most a few micrometres). If a sample enters this range, a sufficient number of molecules in this contact area to the measuring crystal can be stimulated to oscillate, so that the resulting decrease in beam intensity can be detected. The short range of the evanescent waves means that this method is a surface analysis. Because of the advantage that targeted sample areas can be analysed with this method, the ATR method also serves to identify locally occurring impurities. Special µATR techniques (IR microscopy with integrated µATR crystal) allow lateral resolutions of approx. 5 µm to be achieved. Therefore, impurities as well as fillers can be detected locally. This could be helpful to support fault and damage analysis [41, 42, 68–70].

Requirement for sample quality and preparation

Foil or plate segments with dimensions of 30 mm × 30 mm and a maximum thickness of 50 µm are suitable for IR spectroscopic analyses by means of transmission. The requirements differ depending on the equipment manufacturer, so that smaller dimension segments may also be suitable. Plastic flakes from films can in principle be irradiated by means of IR microscopy.

As part of the material identification by means of ATR, it is useful to exclude possible foreign contamination by performing a surface-remote analysis based on the material to be examined. This is realised by first taking a subsample far from the surface (e.g. a thin cut with an area of at least 1 mm × 1 mm). A good contact between the measuring crystal and the sample is needed for material characterisation using ATR. Pressing allows the analysis not only of compact samples but also of powdery and liquid samples. For the detection of fillers and reinforcing materials, the procedure is analogous. In the case of a product that can be optically differentiated, each subcomponent should be exposed preparatively and analysed.

At this point it is important to note that sample residues can remain on the ATR crystal after the sample is removed. Therefore, the measuring crystal has to be regularly checked for such residues and cleaned, if necessary.

In the case of surface analyses, the insertion of further impurities, on the one hand, and the removal of the analyte, on the other hand, must be prevented.

Typical results of failure analysis

Because of the fact that an IR spectrum reflects the stimulation of molecular vibrations, the appearance of detected signals correlates with the complexity of the structural composition: The more diverse the material composition, the more detailed the IR spectrum. This is demonstrated by a comparison of a defective component surface and an inconspicuous surface of a reference sample as shown in Figure 4.16. Only the damage-causing surface shows additional signals that indicate the presence of contamination.

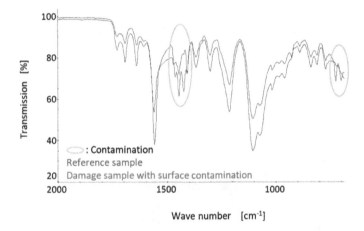

Figure 4.16: Exemplary representation of IR spectra of a damaged-causing contaminated surface compared to an inconspicuous sample surface.

Further examples can be found in chapters 5.3, 5.9, 5.21 and 5.22.

4.2.2 Energy-dispersive X-ray structure analysis (EDX)

Significance for failure analysis

In failure and damage analysis of plastic products, the application of element distribution is particularly useful by spatial resolution, because damage patterns such as material failure can also be associated with a local decrease of reinforcing materials. Furthermore, anomalies obtained by electron microscopy, such as an inclusion, can be characterised and often identified in parallel by EDX structure analysis.

EDX is an elemental analysis method that determines the atomic composition of both organic and inorganic substances. It should be noted that this method does not directly provide structural information, so other methods, such as IR spectroscopy, should be used in addition, especially for organic compounds. An advantage of this element analysis is that semi-quantitative results are possible due to the counting of the signals. A typical application in plastics analysis is the material identification of fillers and reinforcing materials as well as at least the rough material class assignment of contaminants. EDX is also used for the detection of inorganic flame retardants or colour pigments (e.g. titanium dioxide).

In many cases, the overlapping observation of individual elements in a plastic product helps significantly to spatially decompose the mixture into its individual components. This could be helpful to identify impurities in terms of failure analysis. If elements overlap in their positions, this comparison allows detection of not only individual elements but also characteristic groups of substances to be identified in this way. This is illustrated by a nickel–chromium–iron particle, which was detected by EDX in a contaminated plastic sample (see Figure 4.17).

Another example would be the spatial distribution of the flame-retardant filler magnesium hydroxide, which can be evaluated even in the presence of the filler talc (a magnesium silicate) by comparing the distribution of magnesium and silicium. This method also distinguishes filler agglomerates from foreign contaminants.

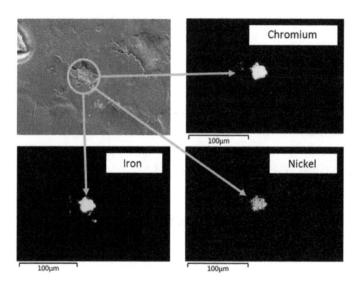

Figure 4.17: Elemental distribution of a spatial overlap of the elements chromium, nickel and iron using SEM/EDX mapping.

Functional principle

Elemental analysis by means of EDX is finally carried out by detecting element-specific X-rays, which are emitted by the material to be analysed. For this to occur, the sample material must first be stimulated. The stimulation occurs already on the part of the imaging procedure, as the sample material is irradiated with an electron beam in SEM. Compared to SEM, elemental analysis is carried out at a defined sample distance from the detector and with a comparatively higher acceleration voltage.

According to the atomic orbital model, different discrete energy levels exist within the electron shell of each atom. If an electron is released that was not at the highest level, there is briefly a free space at an energetically more favourable level for electrons that are located higher up. This space is rapidly filled by an electron from a higher level of this atom. In the course of this, an amount of energy is released in the form of an X-ray quantum according to the difference in energy levels, which is characteristic of each type of atom. This is shown schematically in Figure 4.18.

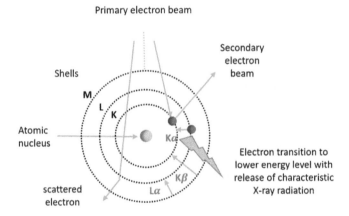

Figure 4.18: Schematic representation of the emission of element-specific X-rays by electron radiation stimulation in the scope of a FESEM analysis.

The result of an EDX analysis shows the signal intensity as the number of detected X-ray photons (pulses as counts per second on the ordinate) as a function of the respective energy of the detected photons (in electron volts; keV on the abscissa). This is shown as an example of the EDX analysis of a filler particle of chalk (calcium carbonate) embedded in a plastic (see Figure 4.19).

In addition, the wavelength of the resulting monochromatic X-ray radiation can even be used to assign which level change (Kα or Lα for example) has taken place because the energy and thus the wavelength depend on the type of electron transition. This explains why several signals are obtained for one element in an EDX spectrum. Using the direct correlation of the signal areas of different elements in one spectrum, the EDX method can also be used for semi-quantitative analysis. In this case, characteristic transitions (dominant level changes of an element) should be used.

At this point it is reminded that the detection of the released X-rays of this method (SEM-EDX) is an effect of the imaging building SE in SEM. Because of the fact that the stimulation source of SEM and EDX are identical (only differ regarding their

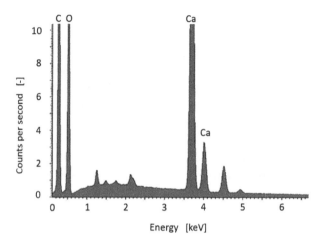

Figure 4.19: Typical EDX spectra of the filler chalk (calcium carbonate) in a polyethylene.

stimulation energy and the correlating penetration depth), there is a good correspondence of the reaction range and thus a penetration depth of a few micrometres, so that the results can be correlated with each other. Therefore, mappings (see below) of SEM and EDX in combination can be used very well for sample assessment. On the other hand, spot EDX analyses are mainly used for the identification of unknown local effects such as inclusion or surface contamination.

By combining EDX and an electron microscope, the respective position and thus the local distribution of detected elements become accessible in an area analysis (so-called mappings). This technique is very helpful to operate failure analysis and can be used to separate different components in a mixture of substances spatially, provided that their elemental composition can be sufficiently distinguished from each other. This is used, for example, to evaluate whether a particular component has been homogeneously integrated. Figure 4.20 shows an example of this in the case of the homo-

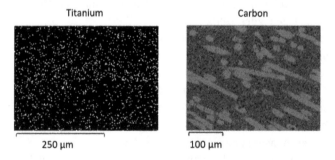

Figure 4.20: Elemental distribution of titanium using SEM/EDX mapping to demonstrate the homogeneous distribution of titanium dioxide in a polypropylene (left) and the presence of the element carbon in a carbon fibre reinforced polyamide (right).

geneous distribution of titanium dioxide when only the spatial distribution of the element titanium is observed (left-hand side) as well as the presence of carbon fibres (right-hand side). The coloured appearances show the presence of the element and the dark areas indicate their absence [51, 74].

Requirement for sample quality and preparation

The requirements for SEM/EDX analysis correspond to those of SEM analysis alone, so it is referred to in Section 4.1.2. With this method combination, it is important to ensure that the sample surface to be analysed is free of any contamination that may have been introduced by preparation. This may require additional cleaning steps. In addition, the use of carbon instead of sputtering with gold can increase the informative quality.

Typical results of failure analysis

A classical EDX spectrum in terms of its element distribution from a point or area analysis can be seen in Figure 4.19. The damage case was a plastic film that showed specks at irregular intervals. Figure 4.21 shows in the upper part of the figure an SEM image with a large inclusion in a film cross-section. The lower half of the figure shows two element distributions of an EDX mapping of that area: The lower left shows the distribution of the element carbon. This element is detected in particular in the foil material of the cross-section analysis. The inclusion itself is based on the element aluminium; the detection can be seen in Figure 4.21 in the lower right half of the figure.

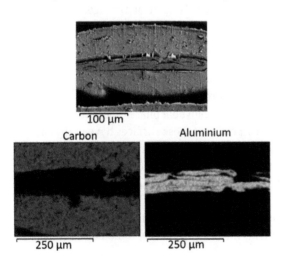

Figure 4.21: Combination of SEM (top) and EDX analysis (bottom) of an aluminium contamination in a polyethylene foil cross-section.

The combination of these methods is often also very suitable in the case of organic mixed systems, as the SEM images can be used to localise the defects and, if necessary, these can be characterised in parallel using EDX.

Further examples can be found in chapters 5.5 and 5.17.

4.2.3 Gel permeation chromatography (GPC)

Significance for failure analysis

Gel permeation chromatography (GPC) is used for the physical separation of molecules that are in solution and in this form inhabit different volumes. The result represents the size distribution of these molecules by plotting the relative abundance against the molar mass. The width of the distribution is called "polydispersity" (PDI). With regard to plastics, GPC can therefore be used to determine the chain length distribution of one or more polymers in a good approximation.

Not only in quality control but also in the field of failure and damage analysis, the average molecular weight distribution by means of GPC analysis is used, for example, to detect possible batch variations, to reveal ageing processes (such as hydrolysis processes or post-cross-linking) and to prove the sample relationship in the case of complaints. If polymer blends are used, the different polymer types can be analysed in parallel, as long as they can be dissolved in the solvent. If the molar mass distributions are of a comparable order of magnitude, signal overlap will be the result. To differentiate between these signals, additional detectors are usually necessary, which operate in particular in the IR, VIS (visual) and/or UV (ultraviolet) range.

Functional principle

The GPC method is based – like any kind of chromatographic separation – on the so-called stationary, immobile phase and a mobile, movable phase. The stationary phase is a microporous filler consisting of many particles whose surfaces have many cavities and pores. These pores are very versatile in terms of size and therefore in terms of both diameter and depth (see Figure 4.22). The separation performance is generated by the wide spectrum of pores, so that this is a basic requirement for the process.

The mobile phase consists of polymer chains that have been successfully transferred into solution and form the statistical cluster with an almost spherical geometry. After the beginning of the measurement, further solvent has to be added constantly during the whole measurement.

In GPC, the separation efficiency is generated by the differing nature of the polymer chains to be analysed, whereby relatively large balls do not or only slightly fit

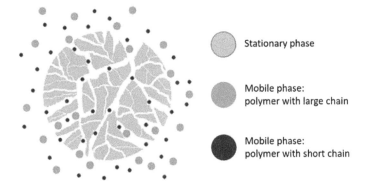

Stationary phase

Mobile phase:
polymer with large chain

Mobile phase:
polymer with short chain

Figure 4.22: Schematic representation of a cross-section of the solid phase to visualise the separation principle.

into the pores of the filler and are therefore pressed through the column quite rapidly. In contrast, comparatively small balls enter also the small channels of the solid phase and thus remain significantly longer in this section of the column. Because of this, comparatively large molecules with higher molar masses leave the separation column first and relatively small molecules last. This is shown schematically in Figure 4.23 using five time-shifted images.

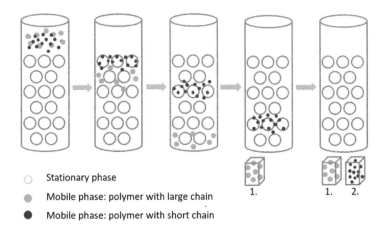

Stationary phase

Mobile phase: polymer with large chain

Mobile phase: polymer with short chain

Figure 4.23: Schematic representation of GPC analysis.

The fractions obtained are analysed with suitable detectors after the column exit, using both concentration detectors (e.g. the detection of the refractive index or UV activity) and molar mass-sensitive detectors (light scattering or viscosity). The resulting data is always based on the data previously specified by calibration, so that each type of detector must be calibrated as well as possible for the type of polymer to be analysed before an analysis is started. For this purpose, several polymer standards with low polydispersity and different mean molar masses are usually used. The molar mass of polymers is typically given in Dalton (Da), which is identical to the atomic mass unit "u" (around 1.66×10^{-27} kg) [71–73].

Requirement for sample quality and preparation

In the case of plastics, the solvent must be selected in such a way that at least the polymeric components can be transferred into solution to behave as mobile phase. In addition to the polymer chains, ingredients such as plasticisers, additives and other substances are also dissolved, but molecules with molar masses below 1,000 g/mol are only detected if necessary. Insoluble components such as fillers and reinforcing materials would permanently damage the solid phase through contamination and must therefore always be removed by filtration before GPC analysis. In this way, polymeric components that could not be dissolved in the solvent used are also separated and accordingly not analysed.

To fasten the dissolving process, temperature increases, enlargement of the surface volume ratio and the movement of the solvent for a faster concentration balance (stirring or ultrasonic bath) can be useful.

Typical results of failure analysis

Within the framework of failure analyses, it is sometimes also necessary to clarify whether the molar mass has changed. This can occur, for example, through hydrolysis processes or through thermos-oxidative degradation. A representative molar mass distribution of plastic with one polymer component can be seen in Figure 4.24. In this case, the molar mass has already been plotted in its weighting in relation to the molar

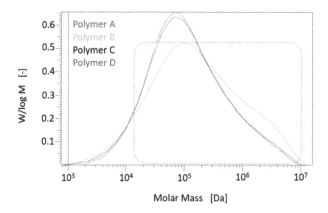

Figure 4.24: Typical molar mass distribution based on a GPC analysis.

mass interval (y axis as W/log M). In this example, four different batches of a product were compared and labelled in a simplified way as polymer A to polymer D. In this case, polymer D provided a deviating behaviour in processing compared to polymers A to C with an additional shoulder in the range of higher molar masses at around 0.3×10^7 Dalton and a lower maximum at around 10^5 Dalton (see marked area in Figure 4.24).

The example in Figure 4.25 shows the relative and thus to the maximum standardised molar mass distributions of three components based on a plastic blend and thus a material with several polymer components. Compared to Figure 4.24, several local maxima were obtained at this point, which could be assigned to different polymer types with the aid of a spectroscopic detector. In the marked area, on the one hand, higher proportions of a low-molecular compound can be observed in the red curves (molar mass range below 1,000 Dalton) and, on the other hand, fluctuations across all samples in the range from 1,000 to 10,000 Dalton.

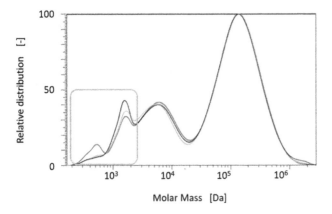

Figure 4.25: Typical molar mass distribution of a polymer blend based on a GPC analysis.

Further examples can be found in chapters 5.2 and 5.21.

4.2.4 Gas chromatography with coupled mass spectrometry (GC-MS)

Significance for failure analysis

In the field of failure and damage analysis, GC-MS coupling supports the observation and evaluation of low-molecular components in particular. Examples include the detection or quantification of additives such as antioxidants or phthalate-based plasticisers as well as the identification of low-molecular contamination. Furthermore, this procedure serves to prove compliance with limit values for maximum permissible concentrations or to provide analytical proof that certain toxic ingredients are not present or in the range of permitted limit values. The evaporation behaviour of, for example, recycled materials can also be investigated and evaluated using a special technique of GC-MS. This allows a qualitative as well as a semi-quantitative determination of the vapours of volatile organic compounds (VOC). On the one hand, it is possible to analyse whether the vapours could pose a health risk and, on the other hand, it is possible to better optimise the processing procedure of, for example, recycled materials by identifying their evaporating VOC as well as controlling the reduction of the VOC amount within the framework of optimisation tests.

Functional principle

As in all chromatographic separation processes, gas chromatography (GC) also separates substances through the interaction of a mobile and a solid phase. The mobile phase is a carrier gas (e.g. helium or hydrogen), which carries the mixture of substances previously transferred into the gas phase past the stationary phase. In GC, the stationary phase is very thin (internal diameter usually less than 0.5 mm) and very long column (usually 10–60 m; up to 200 m are possible in special cases). Independent of the column type, the chemical composition of its coating or filling is important, because the interaction with the sample mixture and thus the final separation performance depend on it. Accordingly, the separation performance and the order of the substances to be separated can essentially be controlled by the polarity of the column coating (non-polar, slightly polar, medium polar and polar) as well as by the diameter and the length of the column and by temperature profile during measurement.

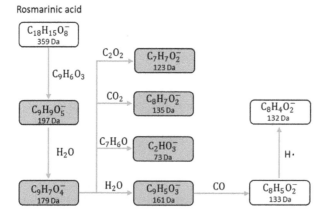

Figure 4.26: Characteristic molecule fragmentation using rosmarinic acid as an example.

The identification of the still gaseous substances emerging from the column with a time delay is carried out by mass spectroscopy (MS) by ionising the molecules chemically or directly by electron impact. In the course of this, molecules are fragmented partially and with reproducible decomposition patterns (see Figure 4.26). The resulting ions are accelerated in the analyser by an electromagnetic field and separated according to their mass-to-charge ratio (m/z) to be finally determined at the detector. During this process, different analysers as well as detectors are used, so that a certain range of differently functioning mass spectrometers is available. Independent of the MS type, molecule-specific mass spectra result, which enable identification through comparison with databases [44, 58, 75–80].

Requirement for sample quality and preparation

Depending on the sample application for the GC, the sample preparation also varies, so that this cannot be generalised. For sample preparation, a basic distinction is made between liquid injection, sampling of a freshly prepared gas phase via the so-called headspace method and the preparation of a gas phase via thermodesorption. In the case of liquid injection, particular attention must be focussed on the choice of solvent, the concentrations of the analytes and the separation of non-volatilisable components such as polymers and fillers and reinforcing materials. In the headspace method, the gas phase is drawn directly from a freshly previously heated sample, so that the temperature and the residence time are of primary importance. If the thermodesorption is too hot, decomposition reactions may occur.

Typical results of failure analysis

In the context of failure analysis, it is not only the polymer component and possible fillers or reinforcing materials that are important with regard to the resulting material properties. In addition, ingredients that are present in small quantities (like additives) can sometimes significantly change their properties. In concrete terms, this means that, for example, the presence of an additive such as a UV/light stabiliser with less than 0.5% by mass determines whether the product is suitable for outdoor use or not.

In a case of defective removal from injection moulded components, it has to be clarified if the amount of release agent was high enough or if the amount was too low (or even a possible absence of this additive) within this remarkable charge which causes the failure. Therefore, both notable components and a sample of the granules were treated with a solvent which is able to solve release agents. Furthermore, granules of a faultless charge (good removal behaviour) were consulted as references. Resulting extracts were analysed by GC-MS. The gas chromatogram of a misbehaving

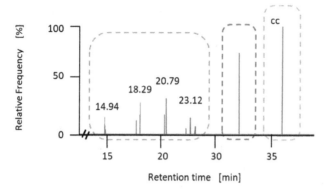

Figure 4.27: Characteristic gas chromatogram of an extracted plasticiser in a PVC tube.

component is shown in Figure 4.27, where the x-axis displays the running time from the start of sample injection up to detection of this substance by MS after leaving the solid phase (column). This running time is called "retention time" in chromatography. It changes depending on general conditions of GC (such as the length and coating of the column as well as measuring temperature). The y-axis shows the relative frequency that sets the highest signal to 100 and normalises the remaining signals accordingly. Since the quantity of the added standard is known, the concentration of all detected compounds can also be calculated via comparison of the signal areas. The red marking indicates the signal of the releasing agent to be evaluated (see also identification by MS below). In green is the added internal standard, which was used to quantify the extracted substances. The signals marked in blue indicate the presence of permitted ketones (not relevant in this case). In the context of the damage case, it

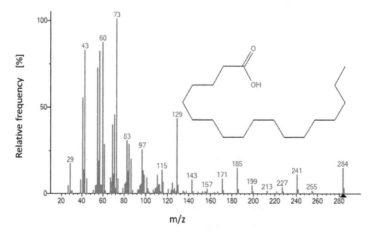

Figure 4.28: Characteristic mass spectra and structural formula of the detected releasing agent (stearic acid) plasticiser DEHP.

was determined that the content of the releasing agent was about ca. 30% lower than in the reference sample, so that the suspicion could be confirmed.

To determine which substance has been detected at which retention time, the qualitative classification is carried out by MS and comparison of resulting fragmentation with the database. The mass spectroscopic identification of the releasing agent by its characteristic fragmentation pattern is shown in Figure 4.28.

4.3 Thermal/physical analysing methods

4.3.1 Differential scanning calorimetry (DSC)

Significance for failure analysis

To obtain information about the material composition at the beginning of damage analysis, DSC provides knowledge about the polymer type as well as details about the material history. This includes the melting and crystallisation behaviour (crystallinity), cross-linking and vulcanisation processes or the stabilisation state (oxidation induction time, OIT/oxidation onset time, OOT). In addition, reheating with previously defined cooling of the same sample material provides helpful information about the thermal history or pure material properties. Another possibility is to detect impurities in the form of foreign polymers and additives (e.g. processing additives) [1, 6, 67, 18].

Functional principle

DIN EN ISO 11357–1 describes two test methods for DSC, which enable the measurement of caloric effects of a sample in comparison to a reference substance. Figure 4.29 illustrates an example of the principle of dynamic heat flow differential calorimetry [18].

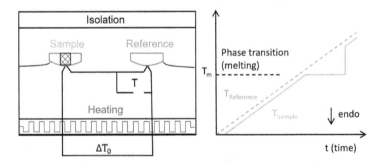

Figure 4.29: Schematic diagram of heat flow DSC/course of sample and reference temperature during a phase transition (melting).

In this setup, the measuring cell consists of an oven where the sample and reference are heated or cooled simultaneously according to a previously defined temperature program. If a physical or chemical heat-tinted phase transition start to occur on the sample side, the temperature in the sample crucible and in the reference crucible differ. The temperature difference results in the relative heat flow, which is proportional to this.

When selecting the temperature program for determining thermal data, the initial temperature should be chosen about 50 °C below the effect to be measured. The final temperature should be a maximum of 20 °C above the effect to be tested; this can prevent aging/decomposition processes. The use of a suitable heating and cooling rate is strongly dependent on the phase transition to be measured. Low rates (for example, 10 °C/min) are suitable for the determination of melting and crystallisation analyses, while higher rates (for instance 20 °C/min) have proven effective for the determination of glass transition temperatures. Usually, the cooling rate should correspond to the heating rate, especially when cooling down to the previously set starting temperature and heating up again. As already mentioned, the first heating serves the thermal prehistory, the second heating the pure material properties. The controlled cooling rate thereby creates a "new" but known sample history. A subsequent evaluation should be based on a comparable evaluation (baseline limits, evaluation method) if possible [67].

Requirement for sample quality and preparation

For a sample comparison, it is essential to prepare the sample at a well-defined location of the specimen. Sampling at the edge or in the core of the specimen has a massive effect on the analysis results. Especially for failure and damage analysis, the choice of sampling is of great importance. If a comparison of a damaged part to a good part or a reference sample is to be analysed, it is advisable to take the samples at the same position of the specimen each time. In this way, misinterpretations can be reduced. The specimens should also be prepared gently (with the aid of a scalpel) to avoid thermal input prior to analysis ergänzen mit [67].

Typical result of failure analysis

A typical DSC result is shown in Figure 4.30 of a semi-crystalline polyethylene terephthalate (PET).

The curve shows the heat flow as a function of temperature. At about 75 °C, a glass transition with a corresponding endothermic effect can be detected. In the further course (at approx. 248 °C) melting of crystalline parts can be shown. By determining the

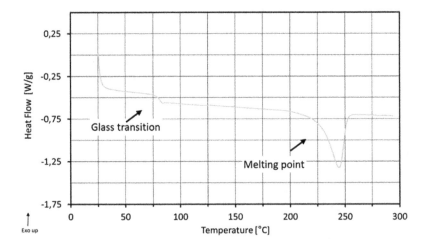

Figure 4.30: Typical DSC result (second heating) using PET as an example.

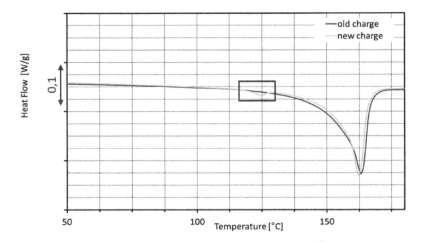

Figure 4.31: DSC curve (second heating) of two PP samples.

area between baseline and peak area, the enthalpy change delta H (crystallinity) during the phase transformation can be calculated.

Figure 4.31 shows the course of two polypropylene (PP) samples. This is an old PP batch that has been in use for years without being stressed. The second sample is a newly used batch from another supplier, which already shows conspicuous behaviour after a short period of use. In addition to the characteristic melting peak of the PP at about 163 °C, a melting of another polymer type (at approx. 125 °C) can also be detected in the new charge. From the melting temperature it can be concluded that this is most likely a high-density polyethylene (PE-HD, see blue marking). Apparently, a recyclate was used in the newly used batch, which meant that in addition to the PP, PE was also present as a proportion of the impurities [1, 67, 18].

Further examples can be found in chapters 5.9, 5.10 and 5.22.

4.3.2 Thermogravimetric analysis (TGA)

Significance for failure analysis

Elevated temperatures lead to thermal decomposition in organic materials. For example, a sustained high temperature in plastics leads to chain breakage, degradation of substituents and oxidation, among other things. These decomposition processes can be investigated with TGA, if they are combined with a gasification. Typical questions in the field of failure analysis are mainly the quantification of the filler contents in the components. Any deviations, for example, in connection with the load-bearing behaviour of the component, can be evaluated by comparative analyses. The advantage of this method over ashing is that the organic additives can be quantified separately from the inorganic fillers and reinforcing materials. In addition, statements can be made about the possible presence of low-molecular substances [1, 67, 19].

Functional principle

TGA is a method for measuring the mass or mass change of a polymer sample as a function of temperature and/or time. The change in mass of the sample occurring during a measurement is compensated for by an electromagnetically or mechanically compensated weighing system by means of optoelectronic sensors to a zero position. From the compensation signal, the mass of the sample is determined as a function of temperature and time. A schematic representation of a horizontal thermobalance is shown in Figure 4.32 (on the left).

The sample is exposed to a dynamic or constant temperature in an oven in the presence of a constant flow of purge gas. The balance (3) records the weight difference resulting from the decomposition. The change in mass can take place in one or several stages, whereby the stages can be assigned to the decomposition of a substance. Ideally, there is a range of constant mass between the stages. In the case of overlapping or closely successive mass changes, the differential measurement signal (derivative, DTG curve) provides further information (see Figure 4.32, right). The example shows a multi-stage degradation. The first degradation stage under an inert atmosphere (in the range of room temperature to 250 °C) shows the evaporation of slightly volatile, low molecular weight substances (moisture, additives and/or monomers). Above 250 °C, typically in the range of 400–600 °C, the (main) degrada-

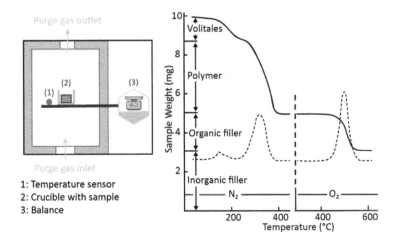

Figure 4.32: Schematic diagram of a horizontal thermobalance/typical measurement result of TGA.

tion phase of the polymer occurs. This can take place in one or more stages (e.g. polymer blends). In the case of plastic samples containing carbon black, switching to an air/oxygen atmosphere can degrade the carbon black and thus provide further information on the material composition [1, 67, 19].

Requirement for sample quality and preparation

The sample weight/sample shape has a significant influence on the onset of decomposition and the peak maximum of the derivative. Therefore, the weights and sample shapes should be chosen equivalently when comparing sample series. According to DIN EN ISO 11358, a sample weight of at least 10 mg is considered reasonable. If comparisons are made at smaller weights, the sample weight should correspond to ±1 mg.

In addition to a suitable location for taking the sample, specimen preparation can usually also influence the measurement result. There is a particular risk with samples that already contain volatile substances. These can already change the initial weight during preparation and set up of the method and falsify the final result. In this case, crucibles with a lid can be used. This is only opened shortly before the measurement starts with the help of a pin or needle and thus prevents premature devolatilisation.

The choice of purge gas can influence the degradation behaviour; if inert gases such as nitrogen are used, the thermal degradation behaviour is injected. If a reactive gas (lust or oxygen) is used, on the other hand, thermo-oxidative degradation is achieved. For example, the degradative mass fraction of soot can be detected in an oxygen atmosphere, whereas this is not possible in a nitrogen atmosphere, where soot does not decompose. This means that the measurement result can be influenced by the change between inert and reactive gases during the test.

Furthermore, the heating speed influences the start of decomposition; a higher heating rate leads to a shift of the start of decomposition to higher temperatures. Likewise, heating up too quickly can lead to strong "overshooting" (briefly exceeding the measurement temperature due to the mass inertia of the furnace) in isothermal tests. Therefore, it is recommended to heat up at a maximum rate of 50 °C/min.

To determine filler contents like the glass fibre content in an injection moulded part, it should be noted that the filler may be unevenly distributed. Because of the low sample weights, a well-founded statement can only be made on the base of a representative sample. An alternative is the determination by means of ignition on larger samples according to DIN EN ISO 1172 [1, 67, 19].

Typical results of failure analysis

A typical measurement result is shown in Figure 4.33. Pump housings made of PA 6 showed cracks, which led to leaks in heating systems. A comparison of a defective component (housing with crack) with a good component made of PA 6 shows differences in the degradation behaviour. First of all, low-molecular substances (presumably different conditioning states) seem to outgas in the sample from the defective component (in the range between 140 °C and 250 °C). Both samples had an almost identical content of inorganic fillers (approx. 30 wt.%).

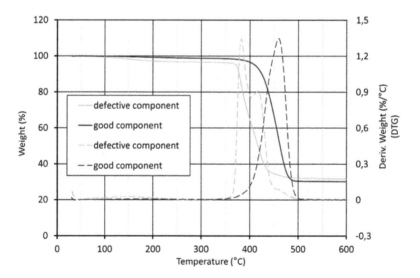

Figure 4.33: Typical measurement result of TGA: TG curve (solid), DTG curve (dashed).

In addition, a significantly earlier onset of degradation could be detected for the sample labelled "defective". This conclusion is supported by the fact that a clear shoulder formation was also detected in the derivation. The early degradation of the polymer is due to molecular degradation (chain degradation), which was also confirmed by supplementary gel permeation chromatography analyses (GPC analyses). The media in the heating systems attacked the material so strongly, which resulted in chain degradation.

4.3.3 Thermomechanical analysis (TMA)

Significance for failure analysis

With the help of the TMA, also known as a "dilatometer", a wide variety of structural changes in the material can be detected. In the field of failure and damage analysis, structural changes in the material, such as material relaxations, orientations and residual stresses, as well as crystallisation and cross-linking processes can be detected. The initial heating of the TMA essentially provides conclusions about the thermal history of the product from moulding to use, the second heating (after controlled cooling) describes similar to DSC analysis more the material characteristics. Furthermore, statements can also be made about the anisotropy in the plastic (component) [1, 67, 20].

Functional principle

By means of the TMA, the one-dimensional thermal expansion is recorded as a function of temperature at constant load. The thermal linear expansion coefficient α can be calculated from the measured expansion of the sample. A schematic measurement setup is shown in Figure 4.34. The sample is placed on a ground quartz glass surface in a tempering furnace and subjected to a constant load with a measuring probe. The sample must have plane-parallel contact surfaces that are perpendicular to the direction of measurement. The sample is then subjected to a temperature programme with a constant heating rate in the furnace (according to ISO11359-1/2) [1, 20].

Requirement of sample quality and preparation

During sample preparation, both the plane-parallelism of the sample surfaces must be ensured and care must be taken to ensure careful sample preparation. Therefore, thermal and mechanical influences (sawing without cooling and clamping in a vice) should be prevented during sample preparation. The place where the sample is taken also plays an important part. If a sample is taken at the edge or in the core or in the

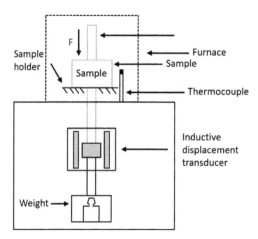

Figure 4.34: Schematic diagrams of TMA aperture.

entire cross-section, the thermal linear expansion is influenced by different cooling conditions and orientations in the component [67].

Different types of quartz glass probes are available for analyses to adjust them to the existing sample geometry (foils, fibres, different cross-sections) and certain questions (coefficient of expansion or measurements of the glass transition temperature). To ensure direct contact between the measuring probe and the sample, a constant, low contact force must be applied during the measurements. To minimise the penetration of the measuring probe into the sample at elevated temperatures, the contact force is kept as low as possible during the measurement [1, 67].

Typical results of failure analysis

Figure 4.35 (on the left-hand side) shows a result of a glass fibre-reinforced PA 66. The measurements were made both in the fibre orientation and perpendicular to the fibre orientations. A comparison of these measurements shows that the expansion in the fibre orientation direction is clearly characterised by the low expansion of the glass fibres, whereas the expansion behaviour of the polymer matrix is predominant perpendicular to the fibres. At a temperature of approx. 52 °C, an increase in the curve slope can be seen, which indicates the glass transition of the PA 66. Figure 4.35 (right) shows the anisotropic behaviour of an ABS injection-moulded component. The change in length of the sample in the injection moulding direction x varies strongly above 90 °C (range of the glass transition temperature) compared to the samples transverse to the injection moulding direction (y) as well as in the thickness direction z. The reason for this different behaviour is that in the x-direction the molecules were strongly oriented and disorientated at increased temperatures [1, 67].

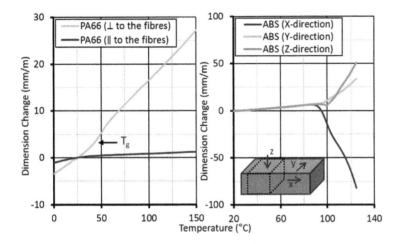

Figure 4.35: TMA of a glass fibre reinforced PA 66 (left)/TMA of an ABS component (right).

4.4 Rheometry

In polymer testing, there is a variety of rheological measurement methods for determining the viscosity of a plastic melt. The following methods are normally used:
- Melt Mass Flow Rate (MFR)
- High Pressure Capillary Rheometer
- Rotating Rheometer

It is important to understand the possibilities and limitations of these various rheological measurement methods and corresponding measuring devices to characterise the specific properties of a polymer. Only then can the various measurement methods for failure analysis be usefully applied and employed [1, 35].

4.4.1 Melt Mass Flow Rate (MFR)

Significance for failure analysis

The value can be found on almost all thermoplastic material data sheets and is often checked during incoming goods inspection. If a variance from the MFR value on the material data sheet is found during the inspection, further investigations must be carried out to determine how this could have been caused. It is important to know that this MFR value does not represent the flow properties during processing, just a point on the viscosity curve at a low shear rate. Furthermore, melt viscosity correlates with the molecular weight of a thermoplastic material, so that the MFR value can be used in failure analysis as an indicator for the degradation of polymer materials or a change in the polymer structure.

Functional principle

The Melt Flow Rate (MFR) value is specified (standard since 1965 DIN EN ISO 1133 and ASTM D-1238). Method B (path length measuring method) of DIN EN ISO 1133 is frequently used. The temperature and weight must first be defined for the material. The material will be melted by the heater after the filling. If the material is complete melted the measurement starts. The weight pushes the melt through the capillary with the help of the stamp. While the measuring distance is passed, the extrudate is collected and the time is stopped. The value is calculated from the measured time and the weight of the extrudate. The functionality is shown in Figure 4.36 [30].

Requirement for sample quality and preparation

Both granules and crushed components can be used as samples. It is important that hydrophilic polymers such as polyamide are dried before measurement. Only a quite small amount of the polymer is required (5 g to 8 g).

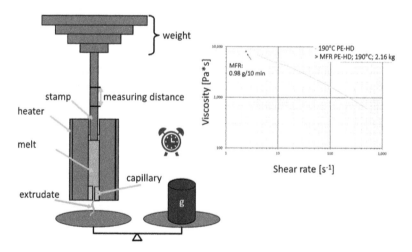

Figure 4.36: Functional principle of MFR measuring method (left) and a typical MFR value in correlation to viscosity curve.

Typical results of failure analysis

The MFR value can be characterised with relatively limited effort, but it characterises the melting behaviour only by a single numerical value (e.g. 0.98 g/10 min at 190 °C with 2.16 kg.; see Figure 4.36. Therefore, MFR is a well-established method used in incoming inspection and for materials that are well known so far. The MFR point in the diagram in Figure 4.36 in the extension of the viscosity curve and is therefore to be considered rather uncritical. If this MFR value is significantly higher (e.g. 3 g/10 min.) compared to the material data sheet, a reduction of the molecular weight is to be expected.

To understand the complex flow behaviour of polymers during the process, the viscosity has to be measured by high-pressure capillary rheometer in a shear rate range and by different temperatures, if possible, in the range of processing parameters, or in a rotating rheometer to characterise the polymer structure by oscillating measurement.

4.4.2 High-pressure Capillary Rheometer

Significance for failure analysis

In plastic processing it is very important to use optimal machine parameters to produce good devices. If anything changes, either in the process or in the material, the production process must be reconfigured. With the High-pressure Capillary Rheometer it is possible to measure the viscosity at temperatures and shear rates as in the processing process. This makes it possible to determine whether the material can be processed in the specified shear rate range at the specified temperatures or to determine if it is anything wrong in the process.

Functional principle

The main advantage of the High-pressure Capillary Rheometer is that shear rheological properties (shear viscosity) can be measured in very high shear rate ranges, as they also appear in plastics processing machines (e.g. extruder or injection moulding machine). First the temperature and the shear rate range has defined for the material. The material will be melted by the heater while and after the filling. It is important to make sure that there are no bubbles in the melt. If the material is complete melted the measurement starts. At different stamp speeds, the melt is pressed through the capillary. While the melt passes the pressure hole, the pressure transducer detects the pressure for the set shear rate close above the capillary.

The extrudate should have a smooth surface and not include bubbles. For each temperature, at minimum two capillaries with the same diameter but different lengths should be measured. The viscosity curves measured in this way are called apparent viscosity curves. For this reason, the inlet pressure loss is then calculated using the Bagley correction, so that a "true" viscosity curve (correct viscosity curve) can now be determined. The functionality is shown in Figure 4.37.

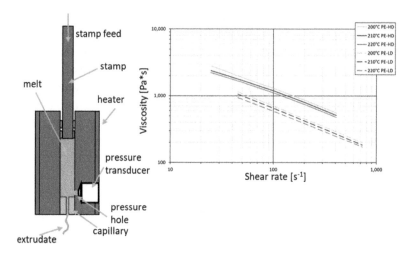

Figure 4.37: Functional principle of high-pressure capillary rheometer.

The diagram shows typical measurement results of two polyethylene types with different densities at different temperatures.

Requirement for sample quality and preparation

Plastics are usually fed to the high-pressure capillary rheometer in granulate form. Flakes shredded from components are also possible. If necessary, care should be taken to ensure that the plastic is well dried. In contrast to the MFR value measurement, significantly more material is required for this measurement (500 g to 1,500 g), depending on the device.

Typical results of failure analysis

An example could be that the material needs to be dried before processing, but the drying unit has a defect. In this case, the incorrect drying has an effect on the viscosity, which is significantly lower than if the material had been dried according to the manufacturer's specifications. The diagram in Figure 4.38 shows a typical viscosity curves at the same temperature and different shear rates of a virgin material compared to measurement curves with significant lower viscosity of the same material (Test_A and Test_B). Causes for these effects can be using of incorrect parameters of the processing machine. In addition, the using of an unqualified material for these machine parameters (e.g. recyclate) can show this viscosity curve. An incorrect drying before the processing can show this phenome for moisture-sensitive polymer. Often a degradation is the result of this observation for these materials [31].

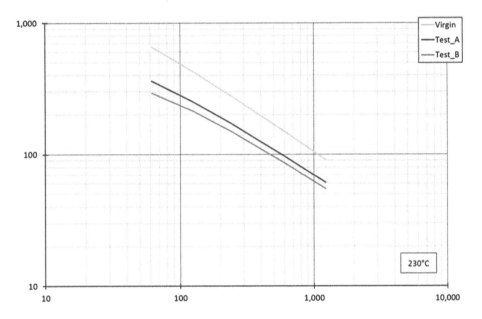

Figure 4.38: Viscosity curve from virgin material vs. two test materials (A and B).

4.4.3 Rotating Rheometer

Significance for failure analysis

A further option to measure viscosity is to use a rotary rheometer. The plate/plate-system with oscillating mode in the melt can be used to get an idea of the molecular structure of the measured thermoplastic material. In most cases, however, the rotation mode is not suitable for use in failure and damage analysis on plastics due to the insufficient torque.

Functional principle

The rotational rheometer essentially consists of two plates arranged concentrically to each other (see Figure 4.39), between which there is a gap into which the plastic sample is introduced in granulate form. This system is surrounded by a furnace that is able to realise temperatures above the melting temperature of plastics. As a result, the plastic sample in the gap is transformed into a molten state and the gap between the plates can be reduced to a gap of approx. 1 mm. Rotating or oscillating movements of the moving plate force the plastic melt to shear. Amplitudes and frequencies are usually specified and the response function of the melt is reproduced in the form of the torque.

In this process, the viscosity determines the torque required for a given combination of angular displacement (angle) and frequency (angular velocity). By varying frequencies (angular velocities) and temperatures, the dependence of viscosity on these external parameters can be determined in this way. The oscillation mode can be used to measure thermoplastic melts. The measurements have to be performed isothermal. As a result, the complex viscosity is given versus frequency or angular velocity. In addition, it is possible to select storage modulus (G') and loss modulus (G") over frequencies or angular velocities out of the data and shown in a diagram. Both diagrams are of particular interest for damage analysis, because indications for the degradation of the material can be generated [34, 36–38].

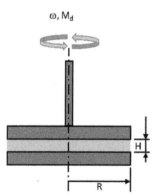

Figure 4.39: Functional principle of plate/plate systems.

Requirement for sample quality and preparation

Both granules and crushed components can be used as samples. If necessary, care should be taken to ensure that the polymer is well dried. Only a quite small amount of the polymer is required (~1 g to ~8 g).

Typical results of failure analysis

The diagram (Figure 4.40) shows a typical curve of the amount of complex viscosity and the cross-over-point of storage and loss modulus at different frequencies of a polyethylene high density (PE-HD). This measurement can be used when, for example, the same material of different batches is to be compared with each other. If the amount of complex viscosity curve is significantly different, e.g. below the expected curve, a degradation of the molecular structure can be assumed.

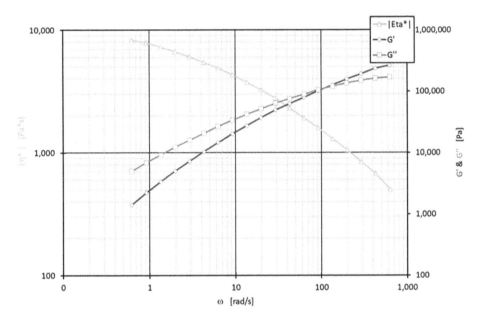

Figure 4.40: Curves of the amount of the complex viscosity, storage and loss modulus.

Further examples can be found in chapters 5.20. For further literature on this chapter see [32].

4.5 Mechanical testing methods

The importance of mechanical test methods in the context of failure and damage analysis lies in the area of verifying the load-bearing capacity of plastic components. If, for example, a component with a load-bearing function is involved, the focus is quickly on load-bearing capacity and operational strength. This is because, especially in the case of recurring dynamic loads or vibrations, the typically listed quasi-static mechanical properties such as Young's modulus or tensile strength are only rough guide values for the load-bearing capacity. The relevant mechanical properties, also depending on the temperature working range of the component, can be characterised in more detail using the following four methods: quasi-static testing, fatigue testing, dynamic cyclic testing and dynamic mechanical analysis.

Requirement for sample quality and preparation

Especially when determining mechanical properties, the positioning of the specimen removal from a component/plate can be decisive: for example, higher mechanical properties can usually be expected in the melt flow direction than transversely due to the stretching and alignment of the molecular chains. A significant influence can also be expected with fibre-reinforced materials, so this must be taken into account when taking test specimens. Furthermore, the manufacturing method of the specimen, the wall thickness, the moisture content and the test temperature as well as sometimes the preparation method influence the mechanical properties.

This effect is intensified, for example, by components contaminated with media, which must also be taken into account during preparation: A container contaminated with oil has usually experienced different filling levels during its life cycle. As a result, there are component areas that had been exposed to a permanent media load (bottom) versus areas with alternating media contact or media-vapour contact. These circumstances must be taken into account in the overall consideration of a sample and should also be considered separately. If a material has been exposed to a permanent media load, it should be considered to determine the material or component characteristics in the media environment or after saturation in the medium. This ensures that the test conditions are similar to the conditions of use. This is especially true for tests with long test durations. Comparative tests should be carried out under the same conditions: sample preparation, test temperature, test parameters. All these preliminary considerations influence the mechanical properties and should therefore be sufficiently taken into account when setting up a test series.

For further literature on this chapter see [43].

4.5.1 Quasi-static testing

Significance for failure analysis

Quasi-static test procedures are carried out with the aid of universal testing machines using different types of loads such as tensile, compression and bending loads. Depending on the specification, realistic operating conditions such as increased/decreased temperature, humidity or medial environment can be simulated. The importance of damage analysis lies, in particular, in the verification of the load-bearing capacity of materials and components, the verification of strength, the investigation of mechanically caused ageing phenomena and the comparison of the mechanical properties of several samples (good and non-good parts).

Functional principle

Quasi-static tests are carried out on universal testing machines according to relevant standards such as tensile test according to DIN EN ISO 527 or a 3-point bending test according to DIN EN ISO 178. The main elements of the universal testing machine are force and displacement/strain measurement technology as well as a crosshead that can be moved at a constant testing or strain speed, which makes it possible to implement a variety of test setups and thus specific load types for standardised test specimens or also components. Figure 4.41 shows the schematic structure of a universal testing machine [25, 26, 66].

As demonstrated in Figure 4.41 a tensile test sample is clamped into a suitable clamping tool. A strain sensor determines the strain of the test specimen in the parallel specimen area. By moving the crosshead in strain control, the test specimen is stretched and the force/deformation curve is recorded. From this, stress/strain curve is calculated where typical properties such as Young's modulus, tensile strength and breaking properties could be determined.

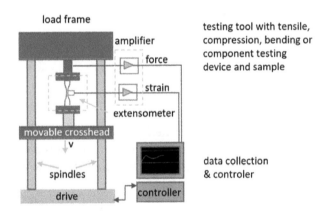

Figure 4.41: Scheme universal testing machine.

Typical results of failure analysis

A possible result of the analysis using quasi-static testing can be, for example, the stress–strain diagram as shown in Figure 4.42. Between 0.05% and 0.25% Young's modulus is calculated [25]. Several factors such as test speed, test temperature, which can be assigned to the test parameters, but also UV-exposure or oil contamination, which can be assigned to the specimen properties, lead to different increase in the stress/strain curve and thus in different results in Young's modulus. In the context of comparative damage analyses, it must be taken into account that only the test parameters can be influenced directly. Therefore, a changing Young's modulus can be an

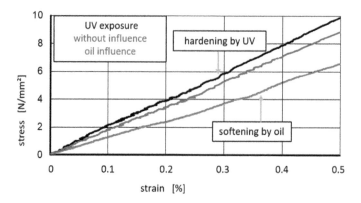

Figure 4.42: Factors influencing mechanical properties of polyamide.

indication of contamination or other environmental influences in the context of a failure analysis. For example, permanent exposure outdoors results in an increase in hardness and Young's modulus. Softening with a decrease in Young's modulus may indicate contamination with media such as oil.

4.5.2 High-speed testing

Significance for failure analysis

High-speed testing is used for mechanically stressed materials and components for which crash-like or high strain rates must be simulated. High-speed testing is also used to adjust high loading speeds and to correlate crash-like fracture patterns with the fracture pattern of failure.

Functional principle

High strain rates and high test energies are realised by using servo-hydraulic machine technology or with the help of pneumatically driven drop towers. In addition to tensile tests, puncture tests and 3-point bending tests under impact loading are important. Using the so-called tie rod, the acceleration of the test axis in tensile tests can be realised with strain rates in the range of up to 100/s or 200/s. The tests are often coupled with high-speed video analysis systems in order, on the one hand, to visualise the failure behaviour for the human eye, which is otherwise often impossible to follow at the speed of the test. On the other hand, grey value patterns on the sample surface are used to analyse local displacements, which allow conclusions about local failure strain. The functional principle of a servo-hydraulically driven high-speed testing machine in tensile mode is shown schematically in Figure 4.43.

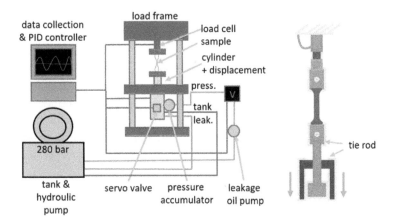

Figure 4.43: Scheme of a high-speed tensile testing machine.

In addition to a load frame with load cell and displacement measuring system, a servo-hydraulic cylinder is connected to a hydraulic system that, being equipped with large pressure accumulators, allows extreme movement of the cylinder. A tie-rod system is used to accelerate the test cylinder, which only takes the specimen along after a certain acceleration distance. The operating of the hydraulic system also includes a PID controller, suitable software as well as the oil tank.

Furthermore, in servo-hydraulically operated high-speed testing machines, pneumatically driven drop impact testers and instrumented impact pendulums offer the possibility of characterising materials under impact loading in ultra-short time. However, the characteristic values determined here, such as energy absorption capacity or impact and notched impact strength, allow less detailed statements about material properties due to the relatively unelaborate test procedure.

Typical results of failure analysis

High-speed tensile tests and slow tensile tests were carried out on PVC roof membranes whose fracture surface showed conspicuously smooth areas in microscopic analysis. By correlating fracture patterns generated in the high-speed tensile test with fracture patterns from the case of damage, it was possible to identify characteristics in common, leading to the assumption that the roofing membranes probably failed under the influence of high loading speeds (storms/wind gusts) in the case of damage.

Compare Figure 5.51 in Chapter 5.23 Cracked roofing sheets.

4.5.3 Fatigue testing

Significance for failure analysis

Whenever materials and especially components or assemblies made of plastics and rubber are exposed to recurrent cyclic loads during their lifetime, then cyclic dynamic tests up to and including fatigue tests must be carried out for complete characterisation. The periodic kind of the load is of great importance in this case, as the dynamic stress limits are generally significantly lower than the quasi-static stress limits. It is therefore not surprising that the fatigue strength is usually less than 40% of the quasi-static strength of the materials. Fatigue properties are carried out with pulsators that are driven electrically, pneumatically or hydrostatically, depending on the load and which can realise tensile, compressive or bending loads.

Functional principle

Dynamic cyclic tests on hydropulsers (Figure 4.44) are carried out using realistic load parameters. By means of a hydraulic system, a servo-valve and corresponding pressure accumulators, as well as in connection with a PID control system, generate a cyclical load on the test cylinder. The force/deformation curves on the test specimen or component are determined by means of load cell and displacement measuring system. In addition, extensometers are used beside the displacement measurement system of the cylinder to be able to observe creep processes at the onset of damage. The repetition frequency of an oscillation is either modelled on the practical application or, if there is a need to reduce the test time, set to a value between typically 3 Hz and

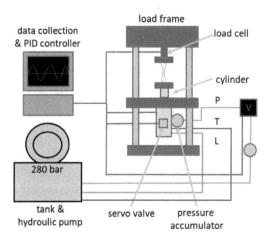

Figure 4.44: Scheme of a hydropulser.

10 Hz at which there is no measurable heating of the test specimen by the imposed oscillation. Loads can be applied in the form of shock-like, ramp-like or harmonic sinusoidal oscillation until a defined target load cycle number is reached.

Generally single-stage fatigue tests are carried out up to a number of load cycles of 1 million or until failure occurs, whereby the number of cycles associated with the load stage is determined and entered in the so-called Wöhler diagram [2, 27, 28].

Typical results of failure analysis

A possible result of fatigue analysis with several load levels can be, for example, the Wöhler diagram as shown in Figure 4.45. Here, the value pairs number of load cycles/ stress from dynamic cyclic tests are plotted according to appropriately staged load levels and connected to form a curve. This diagram can be divided into three areas as specified in DIN 50100 [28] for metallic materials: The first range of up to 10,000 load cycles presents the short-term strength range. This area also includes unique loading cycles such as the quasi-static tensile strength. Between 10,000 and 10 million load cycles, the area is defined as time strength range and above 10 million load cycles, the area of fatigue strength range. The fatigue strength range for plastics is always accompanied by failure (because of blowholes, filler distribution, structure and impurities in the microstructure), so sometimes a statement on the maximum load cycles that can be supported can only be generated up to the actual number of oscillations without fracture. Typically, the fatigue strength, that is, the upper load at which a material just does not fail with infinite repetition of the load, is determined in practical laboratory tests at 10 million load cycles.

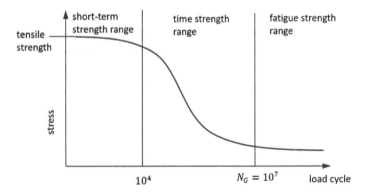

Figure 4.45: Schematic Wöhler diagram of a fatigue test according to DIN 50100, based on data from [29].

For failure analysis, the mechanical fatigue can be determined purely phenomenologically by means of fatigue tests. Often there are indications of the durability of the tested materials and components by the determined number of load cycles. Also, the Wöhler diagram can be used to classify the load level that the material can endure. In this way, further damage as a result of dynamic overload can be prevented.

4.5.4 Dynamic mechanical analysis (DMA)

Significance for failure analysis

Dynamic mechanical analyses in the field of failure and damage analysis are often used in the analysis of characteristic phase transitions and temperature application ranges. Since the softening of a material is coupled with the proportion of crystalline and amorphous areas and their thermal exceedance, the thermal properties are often analysed to prove whether a material is sufficiently mechanically stable/sufficiently strong/stiff in a defined temperature operating range. In particular, because of the filigree design and thus the possibility of testing small geometries, DMA is often used when only small sample sizes or small geometries (for example, parts) are available that cannot be accessed for mechanical characterisation on universal testing machines due to the limited surface area. The use of DMA is advantageous compared to conventional fatigue testing if the temperature dependence of mechanical properties is in focus.

Functional principle

The DMA is electrically driven by a linear motor, which causes a clamp in an oven to move cyclic up and down under low deformations and low forces (a few Newtons). Depending on the test equipment, a tensile/compressive/bending load or shear in the linear elastic deformation range can be realised in the form of shock-like, ramp-like or harmonic sinusoidal oscillation. Different test frequencies can also be analysed. The results are stress or deformation as a function of the test temperature. This allows the analysis of glass transitions, cross-linking processes, influence of fibre reinforcement as well as mechanical parameters such as Young's modulus, shear modulus and damping factor. In addition, temperature application ranges can be obtained from the analyses by means of the temperature ramp applied with the oven [52]. Figure 4.46 shows the DMA schematically in a tension test. The DMA allows mainly two different modes of operation: If a test force is specified, the deformation is analysed as the response; see Figure 4.46 (part 2). If the deformation mode is controlled, the force/stress is analysed as the specimen response. The response of the material is always subject

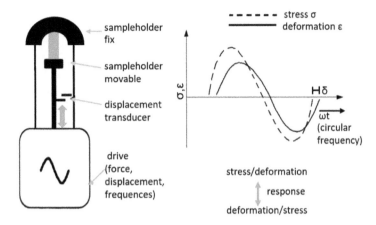

Figure 4.46: Scheme of dynamic mechanical analysator.

to a measureable phase shift, which in turn correlates with the damping properties of the material [1, 2, 52, 66, 67].

Requirement of sample quality and preparation

Similar to the mechanical test methods already presented, aspects such as the position of the sample collection with regard to flow direction, fibre orientation and load direction have to be taken into account. Because of the analysis of thermal properties, special attention must be paid to the thermally gentle production of a plane-parallel sample with a geometry suitable for the load. Further sample pre-treatment such as conditioning may be necessary.

Typical results of failure analysis

A possible result of a dynamic mechanical analysis is the curve of the storage modulus as a function of temperature of three different conditioned polyamide 6 (PA 6). Especially in the field of failure and damage analysis, the temperature operating ranges of polymers specify the operating window of the material for the respective technical application. In the case of PA 6, there can be significant differences in the mechanical behaviour as well as in the temperature operating range due to variations in the humidity ranges. This is illustrated in Figure 4.47. Especially at room temperature conditioned PA 6 has already passed the glass transition temperature so that the segment mobility of the amorphous regions appears. In comparison, dry, freshly injection-moulded PA 6 cannot exhibit such mobility.

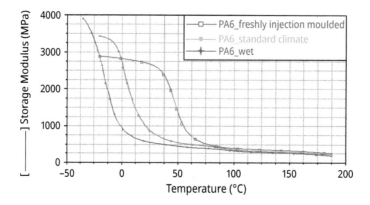

Figure 4.47: Development of the storage modulus of PA 6 under different types of conditioning, based on data from [IKV].

Another example from the field of failure and damage analysis is the characterisation of pallet-like plastic containers made of HDPE. These containers are used for storage and internal transport of, for example, modules. There are two suppliers A and B for the delivery of the transport containers. When testing the containers from supplier B, it turned out that the containers already failed at loads < room temperature. To be able to examine the load-bearing behaviour of both containers in comparison over a wide temperature range, dynamic mechanical analyses were carried out according to Figure 4.48. Here, the supplier A container clearly shows an earlier glass transition area and thus a stronger mobility of the chain segments from about 0 °C. In contrast, the HDPE from

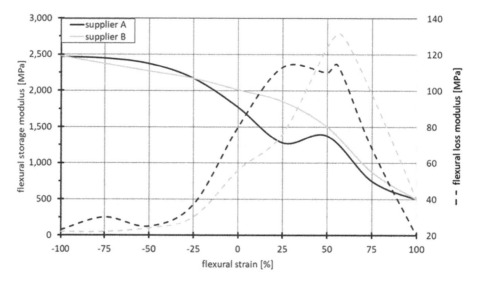

Figure 4.48: Dynamic mechanical analysis of HDPE storage containers.

supplier B has a glass transition at significantly higher temperatures, which means that the chain mobility is very limited at ambient temperatures < room temperature. Here, the glass transition range ends when the 50 °C mark is passed.

In general, the analysed phase transitions at 0 °C to 75 °C are rather untypical for HDPE and it must be assumed that in addition to a main part of HDPE, additional material is present, for example, in the form of a blend, which leads to different mechanical behaviour in the temperature working range. In the example of the material from supplier B, this leads to poorer cold behaviour.

For further literature on this chapter see [5].

4.6 Application of computational methods as a tool for failure analysis and failure prevention

4.6.1 Introduction

In methodological failure analysis, the use of computational methods such as finite element simulations for instrumental analyses has become increasingly established alongside traditional experimental techniques such as imaging technologies to investigate damage patterns using microscopy, spectroscopy or computer tomography. However, simulation methods in engineering originated primarily from the field of component and system design. Their purpose was to ensure structural functionality at an early stage in the development process. Typically, pre-defined requirements are translated into design solutions, which are then verified using simulation methods to determine whether the design meets the specifications and, if necessary, to provide information for adapting the design for a reliable operation. Thus, computational methods are nowadays indispensable tools in the field of failure prevention and can significantly reduce development time and cost by eliminating the number of iterations and prototypes required throughout the development process. In addition to failure prevention, the component overall performance can also be improved by making complex interrelationships manageable with the help of computational methods. For example, in integrative simulation chains, process-induced material properties can be determined in preceding steps and transferred to structural simulations, leading to more accurate predictions for the mechanical behaviour.

Despite the use of advanced computational methods, component and system failures still occur. There are several reasons for this. One is that the real boundary conditions may deviate from those assumed in the development process. The requirements, therefore, do not necessarily correspond to the real conditions. On the other hand, it could be that the chosen models used do not adequately represent the behaviour of the analysed entity. In most cases, however, the calibration effort increases rapidly as the accuracy of the model improves, so there is generally a trade-off between the effort and the accuracy of the computation.

When it comes to using computational methods for failure analysis, the process is analogous to that applied, for example, in component design. However, the requirements and boundary conditions of the computational analysis result from the previous steps of the failure analysis (failure description, failure history recording and failure hypothesis) and not exclusively from the product requirement document. Nevertheless, as discussed before, the ability to capture certain effects and failure modes with computational methods is highly dependent on the models selected and their accuracy in describing the real behaviour of the material and product. Therefore, it is particularly important to investigate the failure mode intensively in the preliminary stages of the failure analysis procedure to choose the right assumptions for selecting the appropriate models to sufficiently predict the behaviour of the analysed component leading to failure. In addition to failure mode and damage characteristics, material properties (morphology, composition, chemical and physical properties), design properties that may affect performance, loads and stresses or processing-related properties need to be considered to provide adequate information to evaluate the failure hypothesis.

Once the modelling depth, model types and numerical methods have been appropriately specified, simulative studies offer a wide range of investigative possibilities, such as parameter studies, sensitivity analyses and the determination and investigation of local properties and states that are not accessible by experimental means. However, because of the large number of applicable models, there is always a risk of errors that can lead to incorrect conclusions regarding the damage hypothesis based on the computationally generated results. Therefore, simulative damage analysis requires both a comprehensive analysis and an accompanying critical examination of the results obtained. However, experimental analysis can also be performed to validate the assumptions, computational results and statements that complement the computational failure analysis. Thus, simulative methods can be a powerful tool for obtaining comprehensive information on the initiation and evolution of failures and for plausibility checks within the methodological approach of failure analysis or for testing specific damage hypotheses.

4.6.2 Principle procedure

Regardless of the type of simulation performed, the approach to the application of computational methods should typically follow a structured and systematic process, which is particularly important in the context of failure analysis. The following is an outline of the essential steps to be taken when using computational methods to perform failure analysis.

Step 1: Define the objective of the computational analysis
It is essential to clearly define what the computational analysis is intended to achieve. The objective is usually derived from the preceding steps of the failure analysis, where the main objective of the computation is often to support or disprove a failure hypothesis. These hypotheses differentiate between several types of errors and are evaluated based on probability and verifiability data from the documentation of the damage and the description of the damage environment.

Step 2: Analysis of the system and pre-processing of the model
This stage involves a comprehensive analysis of the defective component or system. Developing a model based on the information from the previous failure analysis steps should provide deeper insight into the causes and mechanisms of occurring failure. Critical steps include identifying the linkage of the defective component within the overall system and understanding its interactions. This procedure is necessary for defining the boundaries of the computational model, which are essential for accurate predictions. In parallel, the operational and environmental conditions before and at the time of failure need to be investigated and incorporated into the model. Overall, following a previously developed damage hypothesis is advantageous and supports the analysis (see Chapter 2).

Step 3: Verification and validation of the model
This important phase is mainly focussed on ensuring the technical reliability and error-free operation of the built computational model for the failure analysis. The verification is performed to ensure the algorithmic accuracy and computational efficiency of the model to verify that the calculation procedures are in proper agreement.

In addition, the model should be carefully validated to confirm its ability to accurately represent the behaviour of the system under analysis. Model plausibility can also be checked by comparing the model predictions with experimental data obtained from parallel ongoing experimental failure analysis.

Step 4: Performing the computational simulations
In failure analysis, simulations are performed by varying a number of input parameters such as material properties, loading conditions and environmental factors. By varying these parameters, the sensitivity of the system can be assessed, and failure thresholds identified. In addition, simulations can be run under different assumptions, including operating conditions and environmental exposure, allowing different scenarios to be examined, from standard operation to worst-case conditions to be able to gain more information to support or disprove the damage hypothesis.

Step 5: Post-processing and analysis of the results
In this step, the results of the various calculations are extracted and processed. Depending on the depth and scope of the calculations, statistical methods can also be applied to examine the results in detail. The data set generated is then analysed in the context of the failure hypothesis from which information on the cause of the failure

will be determined. Typically, these results are also compared and evaluated with experimental tests.

In addition to the experimental investigation techniques used in failure analysis, computational methods provide a powerful and complementary tool.

For further literature on this chapter see [12].

4.6.3 Application of a computational supported failure analysis

Failure analysis of plastic components is a complex task due to their inherent material characteristics. These components are typically characterised by complex geometries due to functional and part integration. In addition, plastics are multi-scale materials with different morphologies at different scales that significantly affect their physical and mechanical behaviour. Furthermore, these properties are strongly influenced by a wide range of manufacturing parameters, for instance, including polymer flow behaviour, thermal effects and process pressures, adding a further level of complexity to the comprehensive analysis. In this context, the use of advanced simulation techniques, in particular finite element analysis (FEA) and computational fluid dynamics (CFD), become valuable in structural and process analysis. These methods allow detailed modelling of material response under different processing conditions and load

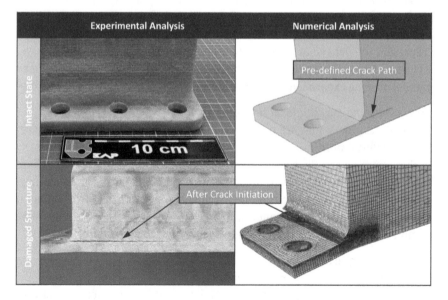

Figure 4.49: Application of numerical fracture mechanics methods for supporting failure analysis in the case of a flange structure made of fibre-reinforced plastic.

cases, enabling the indication of causes for occurring failures and providing an in-depth assessment of mechanisms leading to defective and undesirable component behaviour. As a result, computational methods provide a thorough and detailed framework for failure analysis of defective parts.

In the case of mechanical failure, such as crack initiation, FEA methods are often used to investigate the reasons for damage initiation and progression. Conventional structural design methods generally use continuum mechanical approaches that can predict the occurrence of damage, but do not take into account how a crack is initiated and how it subsequently affects the behaviour of the structure, as the crack itself is not explicitly modelled due to the high computational cost. To investigate crack initiation and progression, fracture mechanics methods must be applied, which require a higher level of modelling. Figure 4.49 illustrates such an application as an example. In this case, crack initiation was observed in a flange structure made of fibre-reinforced plastic. In addition to experimental tests, numerical investigations were carried out to understand the crack initiation and subsequently to identify the critical loads leading to crack propagation. A methodology of numerical fracture mechanics is applied, where the crack path is predetermined due to the given the material composition. This approach allows the initiation of damage to be numerically replicated and the damage hypothesis to be validated.

Rainer Dahlmann, Tobias Conen, Sabine Standfuß-Holthausen,
Christoph Zekorn, Sina Butting, Jan Buir, Meike Robisch,
Michèle Marson-Pahle, Edge Fischer, Christiane Wintgens,
Matthias Klimas

5 Failure and damage analysis examples

This chapter presents typical examples of failure and damage analysis. It is aimed at providing an example of how the process of systematic failure analysis, as described in Chapter 2, is carried out.

Further examples for damage patterns and causes can be found in [2, 8, 9, 11, 13, 17, 23, 39, 53, 60, 61]

5.1 Breakage behaviour despite fibre reinforcement

Problem/damage pattern

An injection-moulded component based on PC/ABS was reinforced with glass fibres to improve its properties. During quality tests, unexpected fracture behaviour occurred, and so the components could not be accepted for their intended use in the automotive sector. The fractured areas are shown in Figure 5.1.

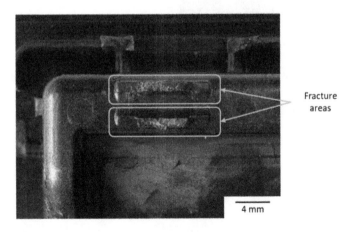

Fracture areas

4 mm

Figure 5.1: Fractured areas of the PC/ABS-GF component.

https://doi.org/10.1515/9783110785647-005

Damage hypotheses

After consultation with the supplier, it became apparent that the raw material had been stored for a long time under humid conditions. It was suspected that the PC had degraded hydrolytically during the manufacturing process (thesis 1). In addition, the component was analysed to identify possible processing faults (thesis 2).

Analyses and interpretation

Because of the assumption of hydrolytic degradation (thesis 1), the average molecular weight distribution was first recorded using GPC. However, by comparing the results of a currently damaged component, the corresponding granulate and an earlier component (without fibre reinforcement), no significant deviations were obtained. Thesis 1 was refuted.

Furthermore, light microscopic analyses were carried out to detect possible causes in the fracture area in the review of thesis 2. Based on cross-sectional analyses in the failure area, processing errors were revealed and thus thesis 2 confirmed: On the one hand, cavities and voids were identified (see dark areas in component curvature in Figure 5.2), which reveal matrix depletion. These represent a significant area of weakness, so that cracks and fractures are encouraged in such areas. On the other hand, a transverse fibre orientation was observed, which indicates that either two melt fronts have collided, or an almost abrupt reorientation of the melt has occurred.

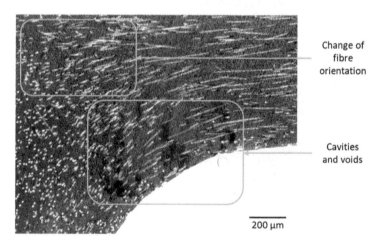

Change of fibre orientation

Cavities and voids

200 μm

Figure 5.2: Light microscope cross-sectional analysis of a defective, glass fibre-reinforced component with the change of fibre orientation and material depletion.

A more detailed examination of the fibres showed that cavities were again present adjacent to the fibres, indicating poor fibre-matrix adhesion. This was confirmed by scanning electron microscopy (SEM) images (right half of Figure 5.3).

Good glass fibre matrix adhesion   Insufficient glass fibre matrix adhesion

Figure 5.3: Comparison of an exemplary good fibre matrix adhesion (left) and an insufficient fibre matrix adhesion in the form of exposed glass fibres with a smooth surface.

Damage sequence, causes, remedial measure

In contrast to the assumption, the polycarbonate part of PC/ABS was not hydrolytically damaged during processing due to possible moisture absorption during storage. Thesis 1 was refuted, while thesis 2 was confirmed by the discovery of several processing errors: The change of preferred direction of glass fibres seems to occur exactly in the failure range, which can lead to premature failure. In addition, there were also cavities in the material precisely in the area of the fracture. A poor fibre matrix adhesion was also observed.

5.2 Component failure: justified complaint?

Problem/damage pattern

In the context of a complaint case, it had to be clarified, on the one hand, whether the complaint samples of single-use reaction vessels originated from the company or whether this common consumer part was manufactured by another company. Furthermore, the cause of the faulty deformation of these products (see Figure 5.4) had to be clarified.

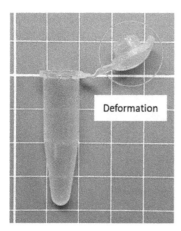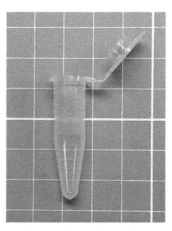

Deformation

Figure 5.4: Comparison of a complaint sample with deformation (left) and a good component (right).

Damage hypotheses

Thesis 1: The complaint is based on a competitor's product. Thesis 2: The product was used beyond the intended application and loads.

Analyses and interpretation

Regarding clarifying thesis 1 material characteristics like polymer type, thermal behaviour, extractable substances and mass distribution were analysed and compared to references based on products from customer itself production as well as from two different competitors with similar products. While some competitor products were able to be distinguished in some analyses, the complaint pattern provided good matches with the reference from the company's production. This is exemplified by molar mass comparison using GPC in Figure 5.5: Comparison of molar masses delivers conformity of the deformed sample only with the molar masses of the customer sample while the values of both competitors do not match with the deformed sample.

To verify thesis 2, adjustment tests using varying thermal loads showed that the adjusted thermal deformation was similar to the damage. This was only observed at temperatures above the specification.

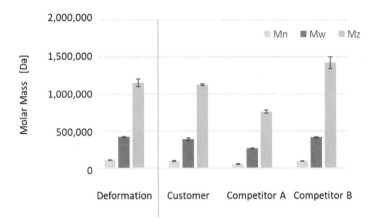

Figure 5.5: Assignment of the manufacturer by molar mass distribution. Conformity of the deformed sample only with the customer.

Damage sequence, causes, remedial measure

Within the scope of these investigations, the deformed single-use reaction vessels show high similarities to the manufacturer's reset patterns. This was revealed by similar average molecular weights, almost identical melting behaviour as well as similar results with regard to extractable substances. The competitor products each show significant deviations from the reclamation samples with regard to the characteristics mentioned. Thesis 1 was thus disproved. Thesis 2 that the defect pattern was caused by external influences that were outside the specification was proven by thermal stress tests. In summary, the complaint was addressed to the correct manufacturer, but was not justified.

5.3 Cracks and delamination

Problem/damage pattern

In the case of a series product in the form of injection-moulded, blue-coloured lids made of PBT, complaints were received from a customer because optical defects had recently appeared at the injection point (see Figure 5.6). The customer described these optical defects as cracks. The cause of these complaints had to be determined.

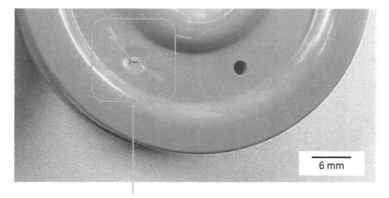

6 mm

Area of cracks and delamination
along the injection point

Figure 5.6: Macroscopic view of the failure pattern.

Damage hypotheses

Based on the screening of the production periods of the complaint reports, it was apparent that all these refer to products that were manufactured with a material modification. Thesis 1: The defect pattern of lids correlates with the alternative, less expensive used masterbatch. Thesis 2: The two masterbatches (low-cost and comparably expensive) can be differentiated.

Analyses and interpretation

Different, blue-coloured lids of two different masterbatches were analysed by light microscopy for a detailed examination of the damaged area to prove thesis 1.

Based on the light microscopy analyses, optical inhomogeneities were observed in both lid types, so insufficient mixing occurs regardless of the raw materials used. This effect is more intensive with the lids of cheap masterbatch. Only these samples also show cracks in the mould area, which are presumably caused by the delamination of the individual layers (see Figure 5.7).

To verify thesis 2, these samples as well as their raw materials were also analysed by infrared spectroscopy regarding material composition. IR spectroscopic analyses deliver polybutylene terephthalate (PBT) as the base material of the lids. While the more expensive masterbatch was compatible with PBT, the new, low-cost masterbatch is based on polyethylene (see Figure 5.8). Compatibility matrices of plastics show that PBT and PE are not compatible with each other.

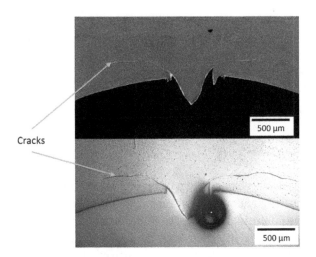

Cracks

Figure 5.7: Transmitted light image (on top) and reflected light image (below) of the lid by the usage of the cheaper masterbatch the cracks in the mould area can be seen.

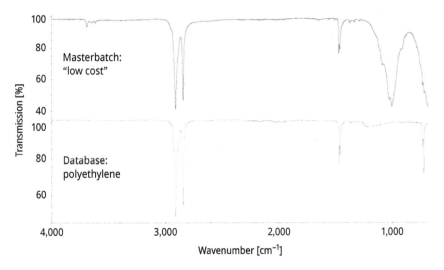

Figure 5.8: IR spectrum of the less expensive masterbatch (upper image area) and the result of the database comparison as a polyethylene (lower image area).

Damage sequence, causes, remedial measure

The light microscope images revealed that inhomogeneities occur when using both masterbatches, so that the process can be improved. However, the lids with the cheaper masterbatch were comparatively of a lower quality, so thesis 1 can still be confirmed.

The material analysis by infrared spectroscopy showed that the cheaper master-batch is based on a different polymer (thesis 2 is confirmed) and is therefore not suitable for PBT.

5.4 Damage causing of water pipe segments

Problem/damage pattern

The subject of the investigation is damaged aluminium composite pipe segments that failed at various points in a hot water circuit of an apartment building after an indefinite period of use (Figure 5.9). The failure manifested itself in the formation of several flaws, which occurred mainly in the axial direction (over several centimetres long) near the hose clamps. The flaws show up as bulges or blisters, some of which are torn open and reveal cracks in the underlying pipe walls.

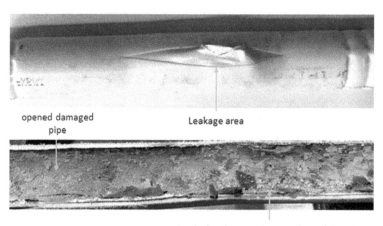

opened damaged
pipe

Leakage area

Cracked and porous inner surface of the pipe

Figure 5.9: Damaged aluminium composite pipe.

Damage hypotheses

The defects present could indicate poor processing quality that can lead to premature failure. Neither the structure of the pipe nor the materials used were known at the beginning. Likewise, the type of material used as well as the state of the material can lead to premature failure. Comparative DSC and OIT (oxidation induction time) analyses were carried out to investigate these questions. The question to be answered was what condition the plastic was in after decades of contact with hot water and what property profile could still be expected after this time.

The investigations were carried out comparatively on the defective component from the hot water circuit and on supposedly intact pipe sections from the cold-water circuit.

Analyses and interpretation

Microscopic examinations of the immediate leakage area revealed that the visible "cracking" in the aluminium was due to the failure of the safety weld. The transverse micrograph (see Figure 5.10) illustrate the composite pipe structure in this context. According to this, the aluminium layer was applied at the time with an overlap of a few millimetres around the plastic inner pipe and joined in a presumably continuous welding process. The cross-section image (Figure 5.11) were taken in the area of a press connector (fitting). Further microscopic analyses also showed that the polymer inner layer predominantly degrades (peeled and porous inner surface) where there was direct water contact.

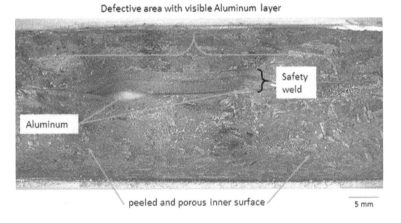

Figure 5.10: Microscopic detail image of the inside of the pipe in the damaged area.

A comparison of the DSC analyses (Figure 5.12, left) between the damaged area (Sample nOK) and an area from the visually intact area (Sample OK) reveals a melting temperature of around 124 °C and 128 °C, indicating that the plastic inner layer is made of PE-HD. The peak shape shows a significantly broader, heterogeneous melting behaviour for the Sample nOK. Furthermore, a higher enthalpy could be detected for the Sample nOK in both heatings. This could have been caused by the decades of use: The amorphous areas of the material exhibit lower long-term resistance to hot water than the crystalline structures and are susceptible to thermo-oxidative chain degradation.

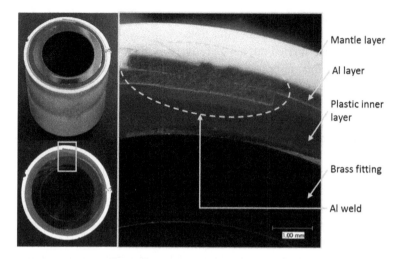

Figure 5.11: Microscopic cross-section analysis in the area of the brass fitting.

Comparative OIT studies from both areas reveal that Sample nOK reacted with the introduced oxygen within 0.2 min, while Sample OK showed a reaction after about 81 min (Figure 5.12). These results suggest that the material's antioxidant protection deteriorated very significantly due to increased temperature in the presence of oxygen, so degradation of the PE has already begun or is imminent.

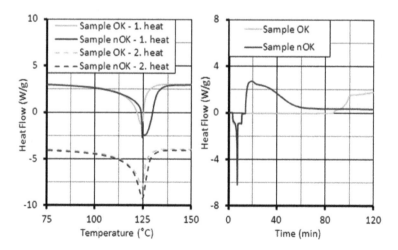

Figure 5.12: OIT results of both pipe segments.

Damage sequence, causes, remedial measure

Overall, the investigations revealed that the examined polymer inner layer (PE-HD) of the hot water pipe segment had lost all its material strength over the service life because of thermo-oxidative aging. The safety weld seam of the aluminium layer was therefore directly exposed to the hot water without protection, which favours led to the failure. Based on the analogous investigations on cold water line segments, it can be assumed that material aging was significantly accelerated by the increased temperature load.

5.5 Investigation of the welding qualities of PP filter frames

Problem/damage pattern

In an exemplary case of damage, injection-moulded filter frames made of black coloured glass fibre-reinforced polypropylene are used as functional elements in paper production. These filter frames were sealed with a polyester fabric. However, one batch showed defects in the sealing process, which led to a failure in further processing (Figure 5.13).

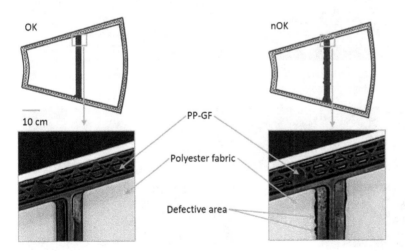

Figure 5.13: Comparison of both filter frame qualities.

Damage hypotheses

The welding behaviour of the filter frames can be influenced by using different polymer types. Likewise, different fillers types or filler contents in the filter frames used can lead to material- or process-specific interactions between the material and the filler.

Analyses and interpretation

During material identification with DSC, clear indications of an additional substance (presumably PE) were found in the damaged filter frame beside the base polymer PP. The investigations on various filler and reinforcing materials (Figure 5.14) revealed considerable differences between the two filters. The filter frame OK contains only glass fibres, while the filter frame nOK contains titanium and calcium as fillers in addition to glass fibres.

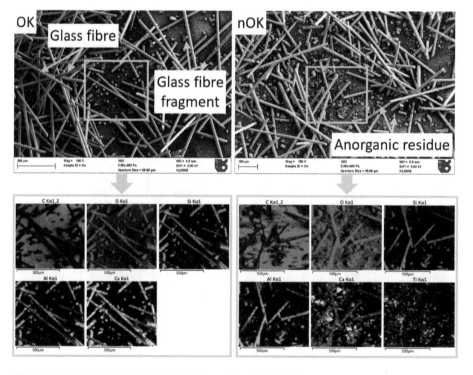

Figure 5.14: Electron microscope pictures/EDX analysis of both filter frames.

Damage sequence, causes, remedial measure

The element analyses revealed that the filter frame of the defective batch also contained chalk and a white pigment. The detection of an additional polymer component and unexpected fillers finally revealed that a PP recyclate had been used in the defective filter frame.

Both the additional fillers and the presence of an additional polymer component in the defective frames have a clear influence on the behaviour in the welding process. Additional polymers lead to a different melting behaviour. Without taking this into account during the welding process, the quality of the seal was significantly reduced. In addition, deviating filler content (or its filler type) leads to an interaction between material and filler. This must accordingly also be taken into account in the parameter selection of the welding process. An increased carbon black content (which was also observed with the thermogravimetric analysis (TGA)) also requires a higher energy input during welding. Without knowledge and ensuring of the specific material composition, it can lead to sealing defects such as flaws, incomplete melting or damage to the material.

5.6 Damage analysis on PP pipes

Problem/damage pattern

About 35 years ago, PP plastic pipelines were installed in an apartment building to convey fresh water. Currently, there have been an increasing number of pipe bursts in the building in question (Figure 5.15). In the course of this, the question currently arises as to whether a precautionary replacement of the old water pipes is necessary.

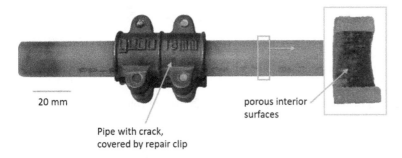

20 mm

Pipe with crack,
covered by repair clip

porous interior
surfaces

Figure 5.15: Damaged pipe section.

Damage hypotheses

The pipe bursts are presumably occurred due to material or processing-related defects in the pipes (foreign inclusions, inhomogeneities). Because of ageing mechanisms, the pipe material was permanently damaged and could no longer withstand the applied pipe pressure (loss of antioxidants, plasticisers). External mechanical damage, for example, during pipe installation, is responsible for the pipe fractures or has promoted them. During an initial visual inspection of the pipe sections, clear colour deviations between damaged and as-new material could be observed, which is why the suspicion of ageing was initially raised. Based on experience, priority was therefore given to verifying the first two damage hypotheses.

Analyses and interpretation

The microscopic examinations show that the damaged pipe sections are interspersed with numerous cracks that initiate on the inner surfaces of the pipes. Some of these cracks extend more than 500 μm into the pipe material (Figure 5.16).

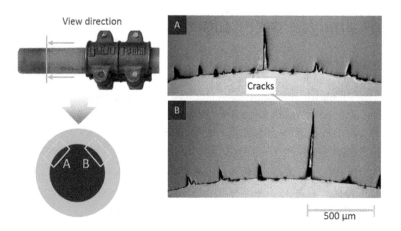

Figure 5.16: Light microscopic sections of the damaged pipe segments.

The inner surfaces of the pipes installed in the building are covered with numerous cracks (Figure 5.17), which is probably the result of material brittleness.

Comparative DSC analyses on the nOK and OK Samples show no significant differences in the melting points or enthalpies in both heating and the associated cooling. The only noticeable characteristic was the early melting of heterogeneous crystal structures. The reason for this could be that the damaged pipe segments had already experienced these temperatures in use. A comparison of the oxidation induction time (OIT) analyses (Figure 5.18) shows that the as-new sample (Sample OK) did not react

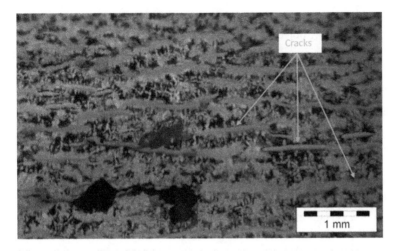

Figure 5.17: Microscopic photography of the inner pipe surfaces/exemplary.

Sample description	OIT [min]	Samples before measurement	Sample description
Sample OK	< 120		
Sample nOK	17.8		

Figure 5.18: Results of the OIT analyses.

over the measured period of 90 min. This is probably attributed to the fact that antioxidants/stabilisers are still present here. In contrast, the damaged samples (Sample nOK) of the pipe segments react with the introduced oxygen within a few minutes. The immediate oxidative degradation could be because antioxidants have been consumed by the operational load of the pipes.

Damage sequence, causes, remedial measure

Based on the investigations carried out, it can be concluded that no significant burst strength can be expected from the pipe sections removed from the building and examined over a longer period of time. Based on the microscopic and thermal analysis results, it can be assumed that material embrittlement has occurred over the period of use by means of aging. Assuming that the pipe sections provided by the client are representative of the building object, it seems advisable to renew the domestic piping system.

5.7 Blistering on electronic connectors

Problem/damage pattern

The term "blister effect" describes the phenomenon of bubble formation on plastic printed circuit board (PCB) connectors, as can occur during automated soldering. The effect is tolerable to a certain extent in some cases. However, if this is accompanied by a major loss of dimensional accuracy, in the worst, case entire assemblies can become rejects. This example deals with such a series defect case, where the housing covers of the assemblies could no longer be mounted correctly due to the blister effect (Figure 5.19). The defective electronic connector was made of PA6-GF30.

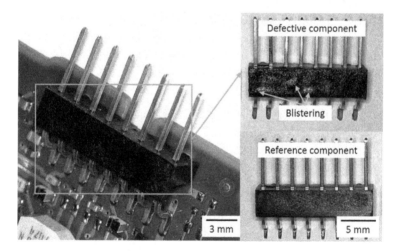

Figure 5.19: Damaged *electronic connector and defective blistering.*

Damage hypotheses

In the course of defective electronic plugs, the main damage hypothesis was that there might be manufacturing-related anomalies. In addition, questions arise as to whether the process temperature was well selected and whether soldering was carried out properly.

Analyses and interpretation

To answer the questions, a macroscopic and microscopic defect pattern documentation was carried out first. Using a thin section preparation of previously embedded plugs, a light microscopic examination of the preparations was carried out in transmitted light.

The affected plugs showed bubble-like bulges on the surface, as shown in Figure 5.20. At higher magnifications, an insufficient surface finish was also visible, indicating that the temperature level of the injection moulding process was too low. The cross-section of the component shows nest-like cavities within the plug, which are blowholes caused by the manufacturing process.

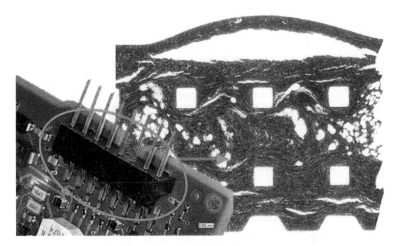

Figure 5.20: Damaged electronic connector and thin section image of this connector.

Damage sequence, causes, remedial measure

Overall, the high-temperature effect during machine soldering caused existing molecular orientations in the component to relax, resulting in the delamination of existing edge layers and thus in the formation of the blistering effect. In the present case, the void fraction in the component was so large that defective connectors could be identified gravimetrically in the production process.

5.8 Cracking due to a lack of chemical resistance

Problem/damage pattern

In waste incineration plants, steel pipes (diameter > 2 m) are located in the area of heat exchangers. These are often lined with protective foils for corrosion protection. The foil is made of perfluoroalkoxy, which is characterised by high permeation barrier and high chemical corrosion resistance.

A defect in the protective foil was detected during a routine inspection (Figure 5.21). The defect refers to detachment of the film from the inner wall of the pipe, expansion phenomena and cracks, which appear to be particularly expressed at the joint welds of various film segments. The foil should be attached to the steel pipe with screws and washers, the heads of which were covered with protective caps (welded foil material). Deformations of the foils in the area of the fasteners indicate that the washer was pulled through the hole.

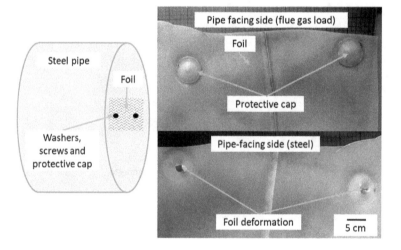

Figure 5.21: Drawing of the steel pipe and defective foil segments.

Damage hypotheses

Thesis (a): Washers were not installed. A lack of these washers would explain why the fixation to the steel pipe was not resistant.

Thesis (b): In addition, inadequate processing quality of the foil and caps is responsible for the damage and a significant source of the cracks and expansion phenomena.

Analyses and interpretation

Thesis (a): The damaged area of the film was examined using macroscopic and laser confocal microscopic methods to identify any notch or imprint made by the washer. Concentric partial impressions could be visualised after opening the caps in the relevant areas, so that the use of washers can be proven (Figure 5.22).

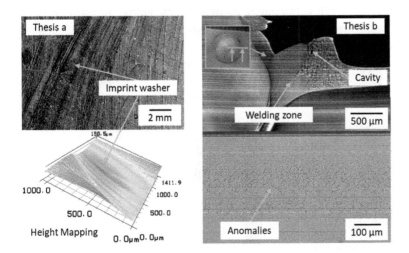

Figure 5.22: (a) Laser confocal microscopic image of the impression of the washer and (b) optical microscopic image of the cross-sections of the film and cap and of the film only.

Thesis (b): A morphological examination of the joint at the protective caps and the film segments of the thermally stressed film shows cavities and notches. In addition, the film shows amorphous appearing edge layers and specimens of the stressed film show further deeper-lying anomalies in the form of a band through the sample cross-section. Presumably, the anomalies are process-related crystallisation effects, which indicate a deviating cooling rate between the specimen core and the edge areas.

Damage sequence, causes, remedial measure

The joints of the protective caps and the foil segments show cavities and notches, which have a negative effect on the stability and corrosion resistance. Because of deformation of the drill hole, the film probably detached from the screw and washer and lost contact with the steel pipe.

5.9 Cracking of chain links

Problem/damage pattern

For several years, metallic chain links have been used for underground mining as cable chains. The chain segments consist of yellow galvanised steel stamped plates, which are encapsulated with polyamide in an injection moulding process. Normally, the component segments are in use for up to 2 years before a wear-related replacement takes place. More recent inspections have now shown only 6 months of service life. The shorter service life was due to brittle fracturing of the polyamide sheaths, as shown in Figure 5.23.

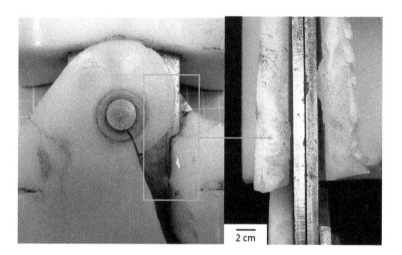

Figure 5.23: Damaged cable chain segment.

Damage hypotheses

The following damage hypotheses can be formulated for the damage case at hand:
a) It is assumed that there was a mix-up of materials and that another material with different properties was used.
b) Because of contamination or too low processing temperature, the processing quality was poor, and the component failed.

Analyses and interpretation

Thesis (a): Investigations by infrared spectroscopy and differential scanning calorimetry were carried out on components with a two-year history of use and as-new components immediately after manufacture to obtain possible material deviations. In combination, the results show that the material is nylon. In addition, similar bands (IR) and curves with identical melting points (DSC) were found when comparing old and new components, so that a material deviation seems unlikely as the cause of damage.

Thesis (b): Fractographic analysis of the fracture surfaces using light and SEM showed that the failure propagates from the bond to the metal insert.

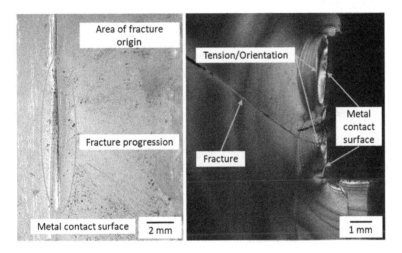

Figure 5.24: Fracture surface of the coating in the area of the metal insert (left) and light microscopic thin section image of the stresses/orientations in the area of the metal insert (right).

Light microscopic microstructure evaluations using the transmitted light method with the addition of polarising filters showed that a comparatively high stress/orientation level is present, which can be observed over the entire component cross-section. However, the feature is comparatively more pronounced in the area of the contact surface to the metal, as shown in Figure 5.24. Because of the manufacturing process, two metal inserts are placed against each other and are over moulded. The microsections show that this results in a largely undefined contour that, from a processing point of view, is not suitable for the product. The overlapping and the sharp burrs of the insert result in offset areas with small radii in the plastic component. In addition, processing-related edge layers appear in the material, which are an indication of a too low process temperature profile. In addition, cold steel inserts were used. Edge layer formation is comparatively more pronounced on the metal contact side.

Damage sequence, causes, remedial measure

The analyses show that in the intended application of the chain links, considerable external stresses act on the plastic, which can lead to melting of the component surface. On the other hand, there are process temperature-related processing deficiencies and construction defaults of the component. In the case of the damage investigated, therefore, the following characteristics probably added up and together led to premature component failure: An insufficiently low process temperature profile resulted in a high residual stress/orientation level in the material. Further stress increased at the contact surface to the metal due to lack of preheating of the inserts (edge layers and rapid heat extraction during over moulding). A notch effect caused by offset or sharp edges in the material (positioning and lack of deburring of the inserts) favours a crack and the accompanying failure of the chain.

5.10 Failure of high-voltage sleeves

Problem/damage pattern

Fuse links are used, for example, in medium-voltage systems to protect the transformers there against short circuits (see Figure 5.25). The fuse links essentially consist of a wound GRP (glass fibre-reinforced plastics) tube containing silver strips with defined cross-sectional tapers wound onto ceramic bodies. The remaining cavities in the GRP tube are filled with a special quartz sand. In the event of an electrical overload of the fuse unit, the silver strips melt in the area of the taper and thus ignite electric arcs. The quartz sand

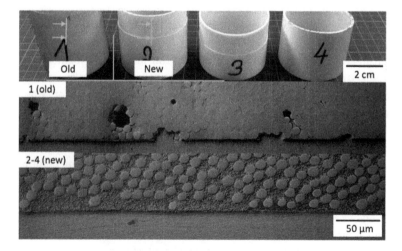

Figure 5.25: Application and cross-section analysis of the fuse body.

acts as an extinguishing medium here, melting locally and extinguishing the arc. The fuse links then have a short-time tube temperature of approximately 200 °C after disconnection. Recently arcs were reignited in internal product tests after electrical disconnection, which then led to the entire fuse burning out.

Damage hypotheses

Differences in the inner life of the fuse links due to production or assembly are ruled out as the cause of the different application behaviour. In particular, the composition of the quartz sand mixture was not changed. Likewise, the silver strips winding and the properties of the fuse body were also unchanged.

For this reason, there is a suspicion that the properties of the tubes, which were purchased from a supplier over many years, have changed, although this was not designated by the supplier.

Analyses and interpretation

In the worst case, scoring or surface material chipping of the GRP cladding tubes could lead to stress peaks and failure. For this reason, the inner surfaces of the tubes were examined under an SEM to identify any differences in roughness or surface-specific deviations in the existing tube qualities. Higher roughness was observed in the damaged GRP pipes. The increase in surface area associated with higher roughness surface area could lead to electrical stress peaks at protruding material tips and therefore have a negative effect on the application.

By means of microsections and a subsequent microscopic examination, the respective tube structure could be depicted in detail. In particular, the wall cross-sections show differences between defective components and earlier production batches. In the case of the damaged specimen, a conspicuously high pore content and a different appearance of the resin matrix can be seen, indicating different fillers, as shown in Figure 5.26.

Analyses using energy-dispersive X-ray spectroscopy show aluminium in the resin matrix of the defective fuse sleeves. Since oxygen was also detected in the interstices, this is probably the filler ATH (aluminium trihydroxide), which is in other applications usually used as a flame retardant. In the case of the defective component, heat led to the splitting off of water and ensured that the current was conducted.

A combination of thermal (DSC) and spectroscopic (IR) analysis should identify any material differences. It was determined that the defective sleeve is based on a phenolic resin, while older sleeves contain polyester systems that also show aromatic structures. In addition, it was confirmed that the flame-retardant ATH was present only in the defective sleeve.

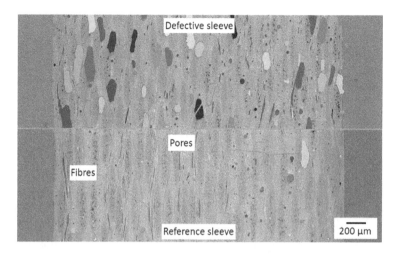

Figure 5.26: Representation of the pore content via the wall thickness of the sleeves.

Damage sequence, causes, remedial measure

Overall, various parameters have promoted the damage that has occurred. On the one hand, completely different materials were used for the damaged components and, on the other hand, processing-related defects in the form of open-pored wall structures contributed to the inadequate behaviour of the component in use.

5.11 Cracking of toilet cisterns

Problem/damage pattern

Toilet cisterns made of ABS have already been successfully produced and sold by customers for many years in a brand-established design. A marketing decision now made it necessary to use a colour masterbatch to change the product colour from natural-coloured ABS to a defined shade of grey. Since the changeover, cracks sporadically appeared in the direction of flow on the cisterns if they were handled roughly during assembly (see Figure 5.27).

Injection moulded toilet cistern   Cracking at the edge
                                            of the housing

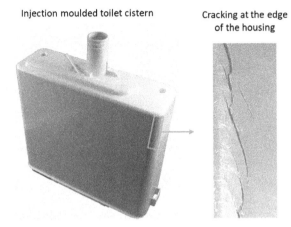

Figure 5.27: Cracking of toilet cisterns.

Damage hypotheses

The fact that the first damage occurred at the same time as the change to masterbatch processing suggests that the cause initially lies in the material or its processing. It is possible that the use of the masterbatch introduced heterogeneities into the product that were not present in the original ABS.

Analyses and interpretation

First, Charpy notched bar impact tests were carried out to determine whether and to what extent the mechanical properties of the grey coloured cisterns were reduced compared to the non-coloured reserve samples. The planar surfaces of the product were ideally suited for the removal of standard-compliant test specimens in an orientation orthogonal to the flow direction. In fact, the notched impact energy of the grey cisterns was reduced by approx. 20% compared to the natural-coloured cisterns.

Further microscopic microstructural analyses were carried out to evaluate the dye dispersion and the associated component homogeneity. A pronounced layer formation in the component appeared (Figure 5.28), which was due to insufficient mixing of the masterbatch with the ABS base material.

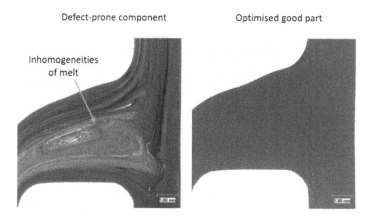

Figure 5.28: Microscopic thin section analysis.

Damage sequence, causes, remedial measure

It is known that layer formations in injection moulded parts can exhibit the effect of interfaces with reduced strengths. Different polymer types show different sensitivities to this effect. Particularly under dynamic mechanical stress, for example, shocks or impacts, such inhomogeneities can become critical to the component, as was the case with the cisterns.

In the present case, it was possible to modify the injection moulding unit by using a static mixing head so that complete homogenisation of ABS and masterbatch was possible. Verification was carried out iteratively via further microstructural analyses, as shown in Figure 5.28. The cracking no longer occurred after the component optimisation.

5.12 Instrument panel breakage

Problem/damage pattern

Dashboards in automotive are often executed as multilayers. In the present case, an injection-moulded carrier made of PC/ABS is laminated with an imitation leather film made of PVC (so-called slush skin). To achieve a soft feel, a polyurethane foam film is also inserted between the PVC and the carrier.

There were isolated customer complaints about these instrument panels from the area of use in the Arab emirates, in which several centimetre long cracks had formed in the PC/ABS carrier after a few months, which were visible as a grade under the PVC skin (see Figure 5.29).

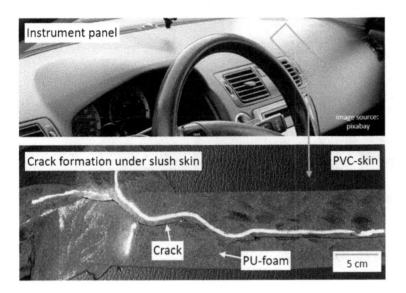

Figure 5.29: Crack formation in automotive instrument panel.

Damage hypotheses

Because of the local limitation of the defect to one area of application of the product established on the market, the suspicion of increased operating stresses (especially high temperature) initially came into focus as a significant influence on the damage. Considering part gating and filling simulations, visual inspection indicated that the damaged area could coincide with a weld. Finally, the formation of a visible edge in the fracture area led to the assumption of increased stresses in the instrument panel (residual, assembly stresses).

Analyses and interpretation

The fracture area covered by the slush skin was exposed and subjected to fractographic analysis to identify the failure origin and fracture type. Further microscopic cross-section analyses were carried out on the material composite to obtain information about the component processing quality (Figure 5.30). Based on intermediate results, further microscope IR analyses were performed, which allowed a chemical line scan in the transition region between the carrier, foam film and slush skin.

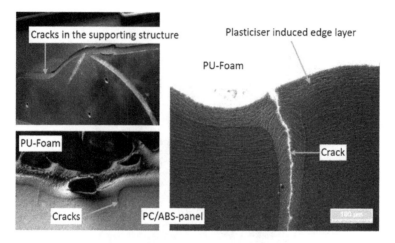

Figure 5.30: Failure and microscopic cross-section analysis.

Damage sequence, causes, remedial measure

Fractographic analysis of the failure surfaces indicated a very brittle material behaviour in the area of the fracture origin, inducing sudden material separation. At the brittle fracture surfaces, the polybutadiene phase contained in the ABS could be evaluated with inconspicuous findings regarding particle size and distribution. The fracture in the carrier started from the contact side to the PUR foam sheet. This was confirmed by the microstructural analyses, which showed numerous small cracks in the panel surface. Another characteristic was an apparently media-induced edge layer in the substrate material, which had also formed around any crack flanks. Investigations by means of microscope IR were finally able to prove that plasticiser migration from the PVC slush skin through the PUR foam sheet into the PC/ABS substrate took place in the course of the increased temperature stress (due to solar radiation approx. 80 °C). This had a strength-reducing effect on the substrate material, particularly in the weld line area, which initiated the crack formation.

5.13 Polyoxymethylene rack breakage

Problem/damage pattern

A toothed rack made of polyoxymethylene (POM) and manufactured by injection moulding is used inside an actuator to convert the rotational movement of a drive motor or gear wheel into a translational movement (Figure 5.31).

This is an established assembly with a successful history on the market. Nevertheless, the manufacturer unexpectedly got confronted with customer complaints according to which the actuators lost their positioning function of the linear actuator despite audible motor operation. An initial inspection or disassembly of the defective parts revealed broken teeth on the rack, which were in frictional contact with the drive pinion during operation.

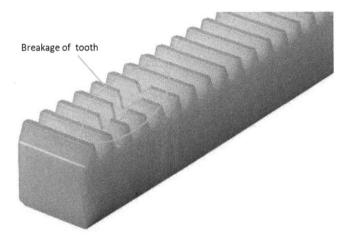

Breakage of tooth

Figure 5.31: Damage pattern in POM gear rack.

Damage hypotheses

Because of the previous successful application history of the product, the design, layout and material selection were initially excluded as possible influencing factors from the case analysis. Since the conditions of use in the assembly could also be assumed to be constant, the component processing initially became the focus of the investigations.

Analyses and interpretation

Microscopic thin-section analyses in polarised transmitted light were carried out to provide information on the microstructure caused by the manufacturing process (Figure 5.32). The investigations revealed indications that the process temperature was too low. In particular, the edge zone in the area of the tooth flanks showed no crystalline structures and was optically separated from the core material.

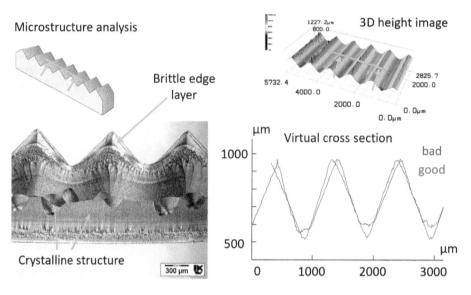

Figure 5.32: Analysis of toothed POM racks.

Damage sequence, causes, remedial measure

If the melt or mould wall temperature is too low, the polymer molecules, which are locally highly stretched by the flow process, are suddenly frozen in the cold edge region, while the slower cooling in the component core provides more time for molecular relaxation and the formation of crystalline superstructures. In the case of POM, this leads to a comparatively brittle edge layer and an unfavourable residual stress profile over the component cross-section.

Furthermore, cold processing conditions during injection moulding can lead to a greater potential for post crystallisation and shrinkage processes.

An improvement can be achieved by increasing the melt and mould wall temperature or using variothermal mould tempering with cooling channels close to the contour. In practice, however, these measures must always be evaluated from an economic point of view.

5.14 Leaks in swimming pool elements

Problem/damage pattern

In the field of swimming pool technology, a so-called basic element made of ABS represents a pool wall bushing and enables the assembly of various installation elements on the pool. The injection moulding of these basic elements is carried out by external companies for the client. The client creates the two-cavity injection mould as well as the design. These basic elements are purchased from intermediaries and pre-assembled before they are distributed to the end customer. Damage in the form of leakage occurred in the swimming pools after 3–9 months of usage. An initial inspection showed cracks near the base element thread, which was adhesively bonded to an installation pipe. Figure 5.33 shows a broken base element with both parts. On the left-hand side of the figure, the fracture surfaces in the area of the stop points are documented in detail. In the middle, microscopic images of the fracture surface can be seen, which show the formation of bubbles. On the right side, the counterpart of the broken base element can be seen, which was adhesively bonded with a black tube from the inside.

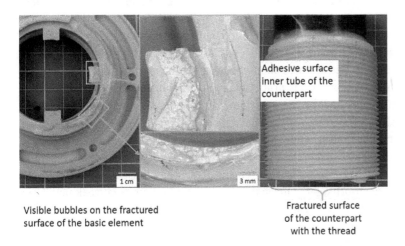

Visible bubbles on the fractured
surface of the basic element

Fractured surface
of the counterpart
with the thread

Figure 5.33: Fractured component and the counterpart with thread detailed light microscope images.

Damage hypotheses

Based on the analysis of the damage pattern and damage environment, various reasons may be responsible for the damage. It is conceivable that processing errors during injection moulding and assembly errors on site are causal. (The manufacturer provides recommendations for the assembly of the components at the end customer's site, which should ensure safe assembly.) Furthermore, unfavourable design properties of the components can promote damage.

Analyses and interpretation

The following analyses were derived from the above-mentioned theses on the root of the damage:

The processing quality of the components was examined, an evaluation of the constructive properties and the design, as well as the examination of possible abnormalities in the failure area. The latter could indicate improper handling during assembly.

The examination of the processing quality was carried out using light microscopic analysis on previously prepared components. In this way, for example, stresses/orientations, inhomogeneities and other conspicuous features induced by the injection moulding process can be visualised and evaluated. Since the fracture is observed in the immediate vicinity of the thread, the focus of the investigations is in this area (thread, area of the spikes and the thread in the valley). Here, a comparison is made between a defective and a new component in the identical component area.

In addition, fractographic analyses were carried out on the fractured base elements with an SEM to find features about the origin of the fracture and its course. On the one hand, this method makes processing-related anomalies/defects such as cavities (blowholes) visible and, on the other hand, it collects possible indications of improper assembly (such as traces of external damage).

In summary, the analyses show that stresses/orientations are particularly present in the area of the spikes inside the thread. Here, the differences in threads geometry seem to have a negative effect (Figure 5.34). The presence of these stress peaks in this area, which has to absorb a large mechanical load, could represent a weak point. The fractographic analysis also revealed that the fracture most likely runs from the inside

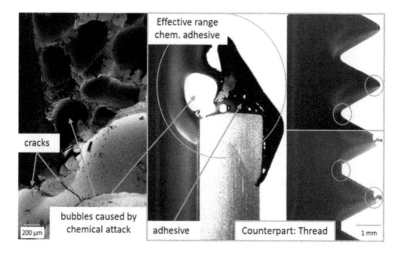

Figure 5.34: Left-hand side and middle: Chemical attack by the adhesive used, right-hand side: comparison of the thread form (new and defective component).

of the thread to the outside. The components show a brittle fracture pattern. It is conceivable that the component design in the area of the thread is not optimal because the surveys and deepening on the inside have a different geometry (between a defective and a new component). On the defective component, these deviations in geometry show higher stresses/orientations. Experience shows that the presence of stress/orientation favours brittle failure.

The presence of bubbles and pores (Figure 5.34, on the left-hand side) on the base element is probably due to the use of an adhesive or other (chemically active) medium. Here, a reduction in component stability is clearly to be expected. The medium (presumably adhesive) seems to have attacked the ABS. In addition to visible blistering, brittleness of the material is to be expected. The light microscopic analyses also show how deep the effective range of the chemically active adhesive is in the components (middle figure). This is also shown by the detailed analyses using SEM in the form of various cracks and incipient cracks on the fracture surface. The interaction of the anomalies already identified and this media attack would have been the cause of the failure of this component.

Damage sequence, causes, remedial measure

The presence of the stresses/orientations detected in the area of the thread, differences in the component geometry in the area of the threads, as well as the use of an unsuitable chemically reactive adhesive, have led to the failure.

The development of a systematic quality test on the basic elements would be recommended. It would be conceivable to apply a defined load or strain in the area of the threads. Also, the adhesives suitable for assembly must be determined and these released to the intermediary.

A re-evaluation of the component design in the area of the thread would also be of great importance at this point. This approach could enable a significant reduction of weak points in the components.

5.15 Cracking of PET bottles

Problem/damage pattern

Cracks in the thread area of refillable PET bottles lead to leaks, causing the contents (mineral water) to spill out. The bottles are a well-established product that has only recently begun to show the cracks mentioned above (Figure 5.35). Prior to the investigations, the customer identified cracks in the lower thread flank area of the PET bottles as the cause of the leak. Within the scope of the investigations, suitable preparation and analysis methods should therefore be used to find indications that can help to clarify the cause of the cracks.

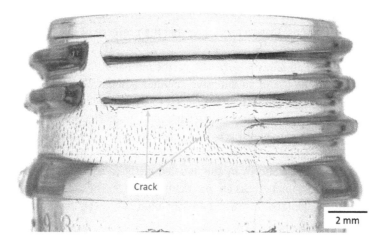

Figure 5.35: Bottle with cracks.

Damage hypotheses

The bottles are disinfected with a washing solution of caustic soda before filling. Because of the described cracks, which are related to sterilisation, the damage occurred due to ESC (environmental stress crack).

Analyses and interpretation

Since ESCs are formed when stresses and a crack-inducing medium are present, light-optical transmissive preparations of the bottles were first made to be able to evaluate their stresses/orientations. By means of polarisation optics, these preparations show processing-related residual stresses or molecular orientations on the basis of the visible interference colours (Figure 5.36). According to qualitative assessment, all bottle specimens examined in the project show a high degree of stresses or orientations in the thread area, which has a detrimental effect on the mechanical properties of the bottles.

On the other hand, pore-like damage to the bottle surfaces in the thread area was found, which was probably due to interactions with the washing solution used (Figure 5.36).

In addition, there are tensile and bending stresses on the bottle thread, which result from the frictional or positive connection with the lid used and which increase the stresses in the thread area.

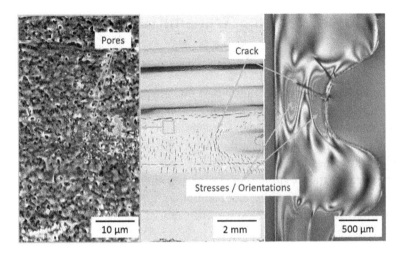

Figure 5.36: Bottle with cracks and pores on the surface.

Damage sequence, causes, remedial measure

The microscopic examinations carried out give reason to believe that the present damage pattern is not subject to monocausality, but that the following features have a damage-promoting effect:
–   increased stresses/orientations in the thread area of the bottles due to manufacturing;
–   lasting interactions between the washing lye used and the PET bottle material (surface roughening);
–   the frictional connection between the cap seal and the bottle mouth can result in deformations of the mouth area, which locally leads to a bending stress of the thread area.

It cannot be assumed that the listed features (occurring individually or in sum) represent a new or individual problem for refillable PET bottles. In the present case, however, the sum of the factors influencing the defect seems to exceed a critical limit, whereby existing pre-damage can grow to an intolerable size over the service life of the refillable bottles.

To avoid the damage-causing ESC in the future, the load collective acting on the bottle thread area must be reduced. Above all, it is possible to influence the manufacturing-related stresses/orientations and the stresses introduced by the cap. This is possible by adapting the manufacturing process and the design of the lidding.

5.16 Failure of an electrical transmitter

Problem/damage pattern

The object of investigation is electrical transmitters that are part of an electrical module (Figure 5.37). These transmitters should be able to emit IR radiation and red light individually or simultaneously, as required. These transmitters, which are only a few millimetres in size, consist of a circuit board on which the two LEDs and two golden cables are welded. The transmitter is then enclosed in an epoxy resin by the manufacturer to protect it from external influences. In this test the electrical resistance of each transmitter is measured, whereby more and more transmitters have recently failed due to an increased electrical resistance. The cause of the increased electrical resistance has to be found.

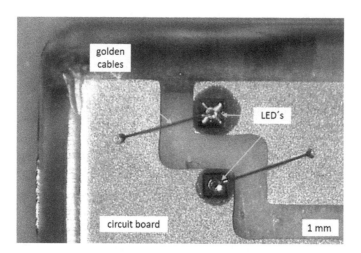

Figure 5.37: Transmitter.

Damage hypotheses

The high measured electrical resistance of the transmitters designated as defective is the result of a conductor path interruption or reduction of the conductor path cross-section.

Analyses and interpretation

First, the transmitters were subjected to a CT analysis to be able to carry out a non-destructive evaluation of the larger conductive paths. No cracks were detected that could be displayed by CT.

Light and electron microscopic images were also taken to visualise interruptions in the conductive paths. These methods were used as a supplement because they have a higher resolution. In the connection area between the LEDs and the gold cables, all the transmitters examined show cracks, which do not separate the entire compound (Figure 5.38).

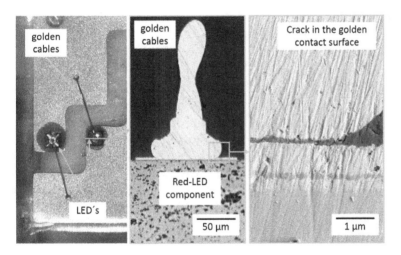

Figure 5.38: Crack in the golden contact surface.

The cracks found are both in the high-electrical resistance transmitters and the low-resistance transmitters. Furthermore, the cracks are always filled with resin. In the direct comparison between transmitters with high- and low-electrical resistance, it was found that the cracks are more pronounced in the transmitters with high resistance (longer and wider cracks).

In addition, the crack formations of the transmitters with high resistance were opened to be able to assess the welding surface. During this preparation, it was found that the welded gold cable had mechanically detached from the LED. The opened cracked surface was examined using EDX analysis. It was shown that these impurities were based on the elements arsenic, gallium, silicon and titanium at different measuring positions. According to the literature, these elements are found in gallium arsenide-based LEDs [24]. It can, therefore, be assumed that material from the LED entered the welding zone when the gold cable was welded to the LED. However, it can be assumed that the electrical conductivity in these contaminated areas is lower than in the gold areas.

The investigations show that a reduction in the conductive cross-section induced by the crack formation led to the electrical defects of the tested transmitters. In addition, impurities were found on the contact surface between the gold cable and the LED, which also led to an increase in electrical resistance. The elements found are components of the LED and lead to the conclusion that the material of the LED melted during spot welding. Since the anomalies found are always within the surrounding resin, a manufacturing-related cause can be assumed. On the basis of the results, the welding process was adjusted, which led to a reduction in the failure rate.

Damage sequence, causes, remedial measure

The investigations show that a reduction of the conductive cross-section induced by the crack formation led to the electrical defects of the tested transmitters. Furthermore, impurities were found on the contact surface between the gold cable and the LED, which also resulted in an increase in electrical resistance. The elements found are components of the LED and allow the conclusion that material of the LED is melted during spot welding. Since the abnormalities found are always within the surrounding resin, a manufacturing-related cause can be assumed. On the basis of the results, the welding process was adjusted, which led to a reduction in the failure rate.

5.17 Flush pipes

Problem/damage pattern

The object of investigation is a black-coloured PP flush pipe bend that is used in concealed WC elements. It is a component that has been proven in use for years (Figure 5.39). The defective specimen originates from a sanitary installation in a private house. Shortly after commissioning, a crack in the flush pipe bend led to water damage in the said residential building.

Damage hypotheses

On the basis of the problem pattern, two theses can be derived. On the one hand, the damage occurred due to incorrect handling during installation of the component. On the other hand, the damage occurred due to poor processing quality.

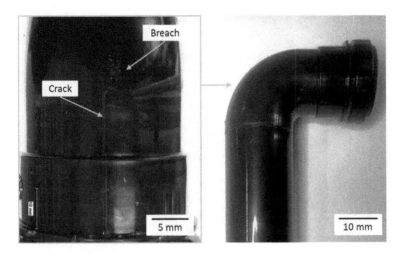

Figure 5.39: Crack on a flushing pipe.

Analyses and interpretation

A fracture surface analysis using an electron microscope was able to characterise the crack surface as a brittle fracture and show that the crack was initiated on the inside of the pipe. On the fracture surface, linear structures can be seen that follow the contour of the component and can be traced back to a layer formation caused by processing. Within this stratification, there is a different form of existing material tips or film formations, which underlines the global microstructure heterogeneity in this case. Particularly at the component edge, a brittle fracture behaviour indicates the existence of processing-related edge layers.

Heterogeneity of the filler distribution can be detected both near the sprue and near the crack origin by means of microstructural examination. Furthermore, with the help of linear polarised transmitted light, strongly pronounced amorphous edge layers can be localised on the inside and outside (Figure 5.40). Heterogeneities of fillers as well as amorphous appearing edge layers are indications of processing-related defects and reduce the mechanical properties of a component.

On the basis of the cut surfaces, inclusions could be detected, which also occur near the crack origin. An EDX analysis of one of these inclusions showed that it consists mainly of the elements iron, oxygen and manganese. For example, this could be tool or screw material that has entered the material through abrasion.

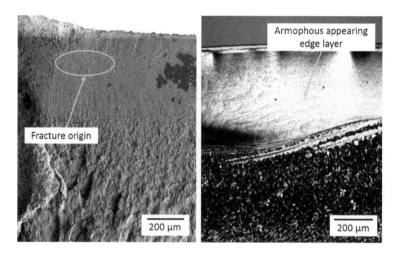

Figure 5.40: Fracture surface and microstructure of the flushing pipe.

Damage sequence, causes, remedial measure

The investigations provided evidence for both theses. The external injuries indicate that there was an impact load on the flush pipe. Because of the punctual distribution of this injury, an impact on a rough surface (such as screed) can be assumed. Since this external injury also correlates with the internal origin of the fracture, further evidence of rough treatment can be found here.

Furthermore, process-related anomalies such as a strongly pronounced amorphous edge layer and metallic impurities were found. Amorphous edge layers reduce the mechanical properties (such as toughness) of a component. Together with the impurities, these anomalies, which can be attributed to the manufacturing process, represent a weakening of the component.

In summary, evidence can be found for both theses. Thus, it can be assumed that the process-related irregularities are the primary cause of the failure of the flushing pipe. It is therefore advisable to change the manufacturing process to a higher process temperature to minimise the pronounced edge layer formation.

5.18 Cracks on circulations pump housings

Problem/damage pattern

Cracks appeared on a circulation pump housing (Figure 5.41), after a short period of use after two weeks. The hosing is made of glass fibre-reinforced polypropylene. The pumps were used in an alkaline process (KOH).

Figure 5.41: Pump housing with cracks.

Damage hypotheses

Poor processing quality reduced the load-bearing capacity of the material. Incompatibility with the alkanic process led to media attack on the pump. As a result, the pump began to crack and leak.

Analyses and interpretation

Microscopic examination of the component showed insufficient processing quality, which should be optimised. This was shown by numerous cavities in the component as well as edge layers. The cavities reduce the cross-section of the component, which means that the component can withstand a low load. An edge layer can lead to lower mechanical stability because it behaves brittle. In addition, a pronounced weld line was found, which was created by the meeting of flow fronts. The weld lines represent a weak point in the mechanical load, whereby cracking can occur preferentially here. It can be observed that the medium has penetrated the material and discoloured it. In this way, the medium could also reach deeper layers in the material. The fracture surface as well as reference components could be examined in the electron microscope, whereby the components with media contact showed clear indications of a media attack. No more glass fibres could be found near the surface, as they had already been dissolved. Deeper in the component, numerous glass fibres were found that had been eroded by the medium. This is shown in Figure 5.42. The reference component without media contact showed no attack of the glass fibres.

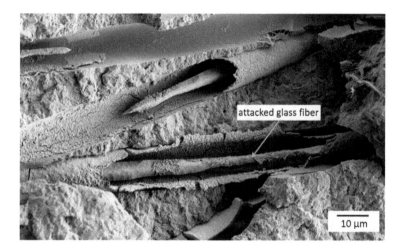

Figure 5.42: Attacked glass fibres (SEM).

Damage sequence, causes, remedial measure

The poor quality of processing has significantly weakened the component from the beginning, as a result of which the voids have reduced the cross-section of the component. The media attack was favoured by exposed fibres and the dissolution of the fibres allowed the media to penetrate deeper. The glass fibres used to reinforce the material are not resistant to the alkali medium.

On the basis of the investigations described above, the following remedy was recommended: To improve the processing quality, the voids, the forming seam and the edge layer formation should be reduced. This can be achieved through various process optimisation measures such as adjusting the mould and melt temperature and a higher holding pressure. In addition, the application of a media-resistant inner coating can protect the component from chemical stress.

5.19 Cracks in post-consumer recyclate tubes (PCR tubes)

Problem/damage pattern

In cosmetic tubes, cracks appeared in about 30% of the tubes after a short storage time (Figure 5.43). These defects occurred only after the changeover to PE (PCR) material. It was noticeable that the cracks always occurred at the same point of the transition from the tube to the tube shoulder. The tube shoulder is injected into the extruded tube, creating an annular gate on the inside of the tube shoulder.

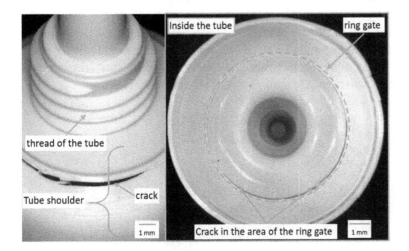

Figure 5.43: Cosmetic tube with crack around the ring gate.

Damage hypotheses

A crack in can indicate a processing defect, insufficient component quality or a possible weakness in the construction/design of the component. Since the construction and design have already been successfully produced with virgin material and no defects occurred, this can be ruled out as a cause at first instance. On the other hand, a previously successful injection moulding process with virgin material cannot be adopted for a PCR material, as the PCR material has other characteristics that must be taken into account. In the case of PCR materials, the processing parameters and batch fluctuations play a major role.

The damage may have been caused by batch variations in PCR quality, making it necessary to adjust the processing conditions. The resulting poor processing conditions lead to locally low mechanical properties and ultimately to the failure of the component.

Analyses and interpretation

Examination of the fracture area in the electron microscope showed that the fracture had occurred in the gate. A macroscopic examination of the gate area showed that this was more pronounced in the PCR material than in the virgin material. This more pronounced gate area of the PCR material acts like a notch and can promote cracking. Further examination of thin sections of the tube in the gate area showed significant differences in the expression of the morphology between the virgin material and the PCR material. The microstructure of the virgin material was more homogeneous. In contrast, the PCR material exhibited more pronounced flow lines and folding of the gate area with cracks, as shown in Figure 5.44. The folding of the gate area indicates early solidification, where each fold may serve as a notch for a crack.

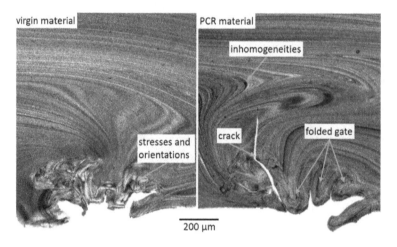

Figure 5.44: Comparison of the gating points in polarised transmitted light.

Damage sequence, causes, remedial measure

The cracks are due to poor processing of the gate in the area of the pipe shoulder. The tube shoulder is injected into the inside of the hose via a ring gate. In this case, there is a solidified weld line with notches in the gate, which serves as the starting point for cracks. In the virgin material, the stresses and orientations are more pronounced (more intense colours), but no cracks appear here. As the quality of PCR material can vary, the processing properties can differ significantly from virgin material and other batches. Accordingly, injection-moulding parameters must always be adapted to the batches and their compositions of the PCR material, and the process parameters must be adjusted in comparison to the virgin material.

5.20 Yellow components (pipes): aging

Problem/damage pattern

Plastic pipes are often used because of their high resistance to various media. The material from which the pipes are made must be equipped with additives to protect the material, depending on the area of application. In this example, pipes made of PE-HD that are installed underground to form long pipelines, were analysed after problems with the welding process. The welding process is used to connect the ends of two pipes. Before welding, the outside surface was removed ("oxide layer"), but the welding process did not proceed as usual. The welding process could only be carried out after a second layer had been removed.

Damage hypotheses

On the left-hand side of Figure 5.45, the first layer removed from the surface is shown. The material is very brittle in comparison to the second layer (right-hand side). The pipes are often stored outdoors until they are installed in the ground. Therefore, it was hypothesised that the material was aged under influence of UV radiation. On the other hand, these pipes do not have a high antioxidant content, because they spend most of their use time underground. This would result in a decrease in molecular weight.

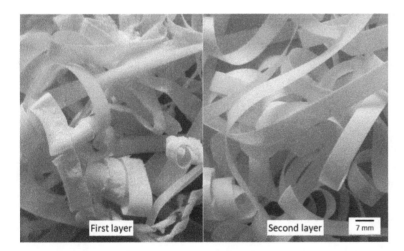

Figure 5.45: Degraded material (first layer) vs. intact materials (second layer).

Analyses and interpretation

The result of an oscillating measurement with rotary rheometer at the same temperature, constant amplitude and different frequencies is shown in Figure 5.46.

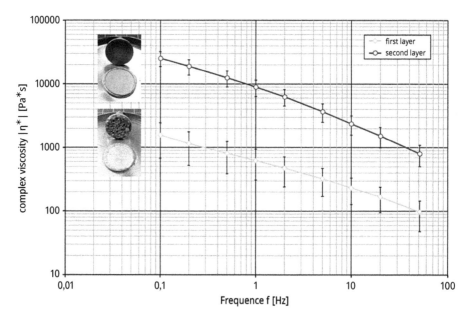

Figure 5.46: Amount of viscosity via different frequencies of the first and second layers.

The viscosity of the first layer is significantly lower than the viscosity of the second layer. A possible cause for the lower viscosity may be a degradation of the molecule structure due to UV radiation.

Damage sequence, causes, remedial measure

Based on the damage symptoms, it was hypothesised that the material on the outside had been damaged by UV rays during outdoor storage. For this purpose, UV additives are normally mixed into the materials. With the help of the viscosity measurements, a difference between the material on the outside and in the core could be proven. This indicates that the thesis put forward could be correct. It is suggested to verify this result with the help of other analyses (e.g. GPC). This did not prove whether enough UV additive was added to the material.

5.21 Fatigue strength of safety-relevant plastic components made of PA6 GF30

Problem/damage pattern

In public transport, a conductor rail is used to supply power to an aerial tramway. The conductor rail is held in pairs and at discrete distances from each other by the so-called support claws. The support claws are injection moulded from a PA6 GF 30 (Figure 5.47). In the present case, a large-scale failure of the support claws occurred.

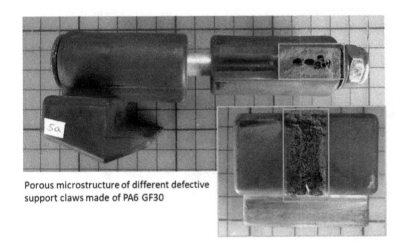

Porous microstructure of different defective support claws made of PA6 GF30

Figure 5.47: Macroscopic documentation of the defective support claws.

Damage hypotheses

It was to be investigated whether this chain reaction could occur due to the ageing of the components and whether other influencing factors could have led to the accumulation of damage. The fracture surface of the support claws, which shows a high number of bubbles/voids (compare Figure 5.47), was classified as conspicuous. In addition to these aspects, the plastic design of the retaining claws was generally considered.

Analyses and interpretation

FTIR analyses as well as gel permeation chromatographic analyses were carried out to investigate the ageing condition of the old support claws. In addition, microscopic analyses and computed tomographic analyses were performed on the retaining claws to evaluate aspects such as fibre-matrix adhesion in failure, fracture behaviour, geometric aspects as well as microstructural quality. This was followed by fatigue tests in the installed condition on a hydropulser to investigate the fatigue strength.

The FTIR analyses confirmed a polyamide as the material for all samples examined and gave no additional indications of oxidation phenomena or aged polymer. This was also confirmed by the GPC analyses, which, taking into account the standard deviation, can be described as inconspicuous.

The microscopic analyses showed a strong agglomeration of voids both for old defective samples and for samples of the new replacement batch (Figure 5.48). All samples reveal a narrow radius in the transition area, which can only poorly compensate for occurring notch effects from the outside. It was observed that the damage originates exactly there and then grows further through the porous structure.

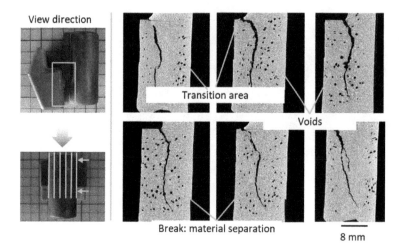

Figure 5.48: Exemplary CT-analysis of an "old" support claw.

The fibre/matrix adhesion, which mainly shows residues of the matrix material on the exposed fibre surfaces, and the fracture behaviour, which is characterised by predominantly ductile behaviour typical of polyamide, are inconspicuous in this context. In the investigation of the fatigue strength of the new batch, all components proved to be fatigue resistant. Marginal increases in displacement after dynamic fatigue loading were documented, but these did not lead to visible damage or to the well-known cracking of the components at the transition area.

Damage sequence, causes, remedial measure

The large increase in the number of damaged areas suggested that an ageing process was the causal factor. However, from the point of view of plastics technology, the support claws show anomalies from the manufacturing process as well as geometric anomalies that have their origin in the design of the component, which are the cause of damage. It is therefore recommended that the support claws be designed and manufactured to suit the plastic. This will increase the load-bearing cross-section and improve the load transfer into the component.

5.22 Porous lamella separator

Problem/damage pattern

Lamella separator systems for flue gas desulphurisation are used for flue gas cleaning. The material used here is polypropylene with 20% talc reinforcement. In flue gas desulphurisation, a gas flow loaded with droplets is guided through a flow grid consisting of curved profiled lamellas (Figure 5.49). In the deflection between the lamellas, inertial forces act on the liquid droplets carried in the gas flow. These forces cause the trajectory of the droplets to deviate from the original gas flow. The liquid film is deposited downwards over the profile wall. The damage to the lamella profile appeared in the area of the upstream side.

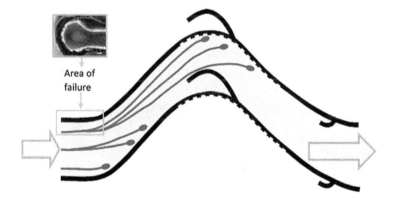

Figure 5.49: The principle of droplet separation in flue gas desulphurisation plants.

During recurring inspections, areas on the upstream side of the lamella profiles were noticed where sedimentation, discolouration or peeling was observed on the upper side of the lamella profiles.

Damage hypothesis

At first inspection, it is unclear whether the surface damage is of superficial sedimentation or whether the material is actually delaminating. Peeling and delaminating layers of material would result in a reduction of the load-bearing capacity. It is noticeable that neighbouring components of the plant do not show any damage (same material, different supplier).

Within the scope of the analyses, two damage hypotheses were initially pursued: The damage is caused by foreign sedimentation or material erosion/corrosion. Errors in processing or in the material composition could lead to early material erosion. Exposure to fumes and prolonged thermal/medial stress could lead to use-related foreign sedimentation.

Analyses and interpretation

The result of a DSC analysis shows comparable melting points at approx. 158 °C to 162 °C for the surface layer and core material. Compared with literature values, this initially speaks for a PP that usually has melting points of between 160 °C and 165 °C. The infrared spectroscopic analysis initially shows a typical spectrum of a talc-reinforced polypropylene. Analyses of the surfaces of the core and surface layer material showed oxidation phenomena in some areas. Oxidation results in a mixture of different degradation products such as carbonyl, carboxyl, alcohol and ester products. These oxidation phenomena were not detectable on the cut surface of the core material, so it can be assumed that the corrosion occurs from the outside and has not pene-

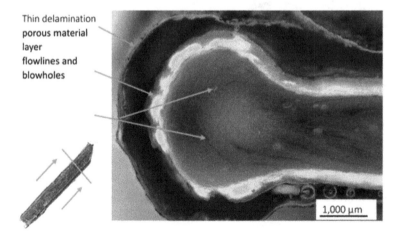

Thin delamination
porous material
layer
flowlines and
blowholes

1,000 µm

Figure 5.50: Microscopy analysis of PP-profile in thickness direction in the area of failure.

trated the entire cross-section. The TGA detected a talc content of 20% and confirmed the specified composition of the material.

Light microscopic analyses were used to show the processing quality and the morphology of the lamellae in cross-sections. Figure 5.50 shows a selected area in the polished section that was in contact with the combustion and cleaning media on all sides. The image shows in the outer area the already delaminating, pergament-like edge layer. Below this peeling, a heavily damaged layer of material is visible, the consistency of which appears porous. This is followed by the core material, which shows an unusually strong yellow colouration. In addition, the morphology of the sample shows strong layering, flow fronts and bubbles. The flow fronts were probably due to poor processing quality. The formation of these flow lines occurs, among other things, when warm and cold melt fronts lie next to each other during processing as a result of an insufficiently homogenised melt. The boundary layers of these flow fronts represent an increased possibility of penetration by media and also reduce the mechanical properties – especially perpendicular to the layer formation. The formation of bubbles may be caused by moisture in the material as well as by insufficient degassing. In addition, a strongly pronounced edge layer, caused by a mould temperature that is too low, may tend to peel off over time.

Damage sequence, causes, remedial measure

Component sections with medial contact showed damage depths of 320 µm on average as well as oxidation phenomena. In addition, the examined sections showed a poor processing quality. The cause of the damage is therefore the medial load, which causes oxidative attack. The processing faults have a damaging effect as well. Remedial action could be taken, for example, by adding antioxidants and optimising the processing quality.

5.23 Cracked roofing sheets

Problem/damage pattern

A large-scale damage to a PVC roof covering a large-volume hall with conspicuously fissured areas of the roofing membrane no longer had any impermeability after the damage and had to be replaced (Figure 5.51). Because of an insurance claim, it was to be clarified which conditions had been the cause of the extensive damage.

Figure 5.51: Damage pattern of cracked roofing sheets.

Damage hypothesis

Analyses by means of SEM showed conspicuous features in the fracture pattern, in which extremely smooth fracture areas were present, which are rather unusual for the PVC material used. Extremely smooth PVC fracture surfaces and very low amounts of ductility could have been caused due to extreme temperatures or extreme deformation/ loading speeds. Because of normal temperature conditions at the time of damage, the thesis of extreme loading speeds, for example, due to storm, is pursued. For verification, simulating tests with the help of mechanical testing machines can be used. For the simulation of extremely high and normal loading speeds, tensile tests on PVC cut-outs were carried out on a servo-hydraulic high-speed testing machine and a quasi-static testing machine. Subsequently, the fracture patterns thus created at different testing speeds were analysed for similarities and differences by means of SEM.

Analyses and interpretation

Quasi-static tensile tests on a universal testing machine and high-speed tensile tests on a servo-hydraulic high-speed testing machine were carried out using tensile test specimens punched out of the roofing film in accordance with ISO 527. Subsequently, the fracture surfaces produced were analysed by means of SEM and the generated fracture surfaces were compared with fracture surfaces from the actual damage case (Figure 5.52).

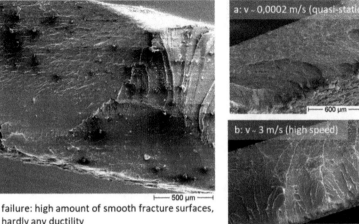

a: v ~ 0,0002 m/s (quasi-static)

|—— 600 µm ——|

b: v ~ 3 m/s (high speed)

|—— 600 µm ——|

|—— 500 µm ——|

failure: high amount of smooth fracture surfaces, hardly any ductility

Figure 5.52: Fracture analysis after high-speed testing.

Damage sequence, causes, remedial measure

The comparison of the artificially generated fracture patterns with the fracture pattern from the damage case showed a clear agreement with the fracture pattern from high failure speeds, since the smooth fracture areas predominate here as well, as found in the damage case. Specimens with quasi-static tensile loading showed remarkably wide areas of ductility (filaments, plastic deformations) in the fracture surface. This suggests that the PVC roof covering was destroyed due to high loading speeds, such as during storms.

5.24 Hailstorm on multi-skin sheets

Problem/damage pattern

Multi-skin sheets are used on industrial roofs. The polycarbonate sheets have been in use for more than 15 years and show numerous holes between the webs, which are limited to the exterior surface exposed to the weather (Figure 5.53). On behalf of an expert appointed by an insurance company, it was to be analytically clarified what the hole formations are caused by. Although the insured event was reported as hail damage, the average size of the holes initially made it seem unreasonable to assume that such small hailstones could cause such hole formation, taking into account standard hail impact tables.

Figure 5.53: Hole formation in roof multi-skin sheets.

Damage hypotheses

The industrial roof affected by pitting was inspected during a site visit to evaluate the local conditions and collect evidence. Because of the inaccessibility, local distribution and frequency of the hole formations, vandalism (e.g. shelling) could be excluded as the cause. No projectile-like objects were found in this context. However, the roof surface also did not show any other characteristics beyond the holes commonly found in hail damage (e.g. dents in zinc cladding).

Analyses and interpretation

Adjustment tests in the form of instrumented hail shots were carried out to show whether and under which general conditions the present hole formations can be caused by hailstones From this, it could be deduced that smaller projectiles had to be responsible for the hole formation on the damaged panels Preliminary mathematical considerations led to the result that 9 mm hailstones can be expected to have an impact energy of only 0.03 J at max. 13 m/s. Based on the common Torro scale, such small projectiles on plastic roofs do not cause any damage yet. However, in adjustment tests, it was possible to generate penetrations with a 9 mm PA projectile velocity of 12.4 m/s and 0.03 J by means of shooting test. The holes created were comparable to the authentic ones. It could thus be shown that 9 mm hailstones have sufficient impact energy to penetrate these multi-skin sheets in their actual condition. Further microscopic examinations revealed that the multiwall sheets had lost approx. 20% of their outer wall

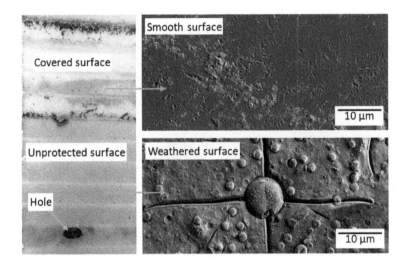

Figure 5.54: Analysis of PC-panels.

thickness due to weathering. In this correlation, polybutadiene particles that had been added to the PC to increase the impact strength were exposed (Figure 5.54).

Damage sequence, causes, remedial measure

It could be proven that as-new multi-skin sheets in the given design can easily withstand a hailstorm under the boundary conditions discussed. However, weather-related aging mechanisms are given too little importance in the damage classification according to customary industry standards at this point. Overall, these led to a significant reduction in the mechanical properties, so that even small-diameter hailstones were able to form impact-induced holes.

References

[1] Dahlmann, R.; Haberstroh, E.; Menges, G., Menges Werkstoffkunde Kunststoffe. Vol. 7. 2021: Hanser Verlag

[2] Ehrenstein, G.W.; Kunststoff-Schadensanalyse. 1992: Hanser Verlag

[3] Hertzberg, R.W. and Manson, J.A., Fatigue of Engineering Plastics. 1980: Academic Press

[4] Kaiser, W., Kunststoffchemie für Ingenieure. Vol. 3. 2011: Hanser Verlag

[5] Naranjo, A., del Pilar Noriega, E.M., Roldan-Alzate, A., Sierra, J.D., Plastics Testing and Characterization. 2008: Hanser Verlag

[6] Höhne, G.W.H., Hemminger, W.F., Flammersheim, H.-J., Differential Scanning Calorimetry. Vol. 2. 2003: Springer

[7] Woodward, A. E., Atlas of Polymer Morphology. 1989: Hanser Verlag

[8] Kurr, F., Handbook of Plastics Failure Analysis. 2015: Hanser Verlag

[9] del Pilar Noriega, E., M., Rauwendaal, Ch., Troubleshooting the Extrusion Process. Vol. 2. 2010: Hanser Verlag

[10] Kinloch, A.J., Young, R.J., Fracture Behaviour of Polymers. 1983: Applied Science Publishers London and New York

[11] Engel, L., Klingele, H., Ehrenstein, G., Schaper, H., Scanning Electron Microscopy of Plastics Failure. 2010: Hanser Verlag

[12] Kämpf, G. Industrielle Methoden der Kunststoff-Charakterisierung. 1996: Hanser Verlag

[13] Bichler, M. Kunststoffteile fehlerfrei spritzgießen. 1999: Hüthig

[14] Schnabel, W. Polymer Degradation, Principles and Practical Applications. 1981: Hanser Verlag

[15] Sawyer, L. C., Grubb, D. T. Polymer Microscopy. 1987: Chapmann and Hall Ltd.

[16] Ehrenstein, G.W. Schadensanalyse an Kunststoff-Formteilen. 1981: VDI-Verlag

[17] Ezrin, M., Plastics Failure Guide. 1996: Hanser Verlag

[18] DIN EN ISO 1357-1. Plastics - Differential scanning calorimetry (DSC) - Part 1: General principles (ISO 11357-1:2016), 2016: DIN Deutsches Institut für Normung e. V.

[19] DIN EN ISO 11358-1. Plastics - Thermogravimetry (TG) of polymers - Part 1: General principles (ISO 11358-1:2014) 2014: DIN Deutsches Institut für Normung e. V.

[20] ISO 11359-1. Thermomechanical analysis (TMA) - Part 1: General principles: The International Organization for Standardization (ISO)

[21] Ehrenstein, G.W., Polymeric Materials. 2001: Hanser Verlag

[22] Hopmann C., Michaeli W., Einführung in die Kunststoffverarbeitung. Vol. 8. 2017: Hanser Verlag

[23] VDI 3822 Blatt 2.1.2 - Failure analysis - Defects of thermoplastic products made of plastics caused by faulty processing. 2012: Beuth Verlag

[24] Globisch, S., Lehrbuch Mikrotechnologie. 2011: Hanser Verlag

[25] ISO 527. Plastics – Determination of tensile properties. 2019: DIN Deutsches Institut für Normung e. V.

[26] ISO 178. Plastics – Determination of flexural properties. 2019: DIN Deutsches Institut für Normung e. V.

[27] DIN 53442. Plastics – Flexural fatigue testing of plastics. 1990: DIN Deutsches Institut für Normung e. V.

[28] DIN 50100. Load controlled fatigue testing – Execution and evaluation of cyclic tests at constant load amplitudes on metallic specimens and components. 2016: DIN Deutsches Institut für Normung e. V.

[29] Broeckmann, C.; Beiss, P., Werkstoffkunde 1, 2013: Mainz-Verlag

[30] DIN EN ISO 1133-1:2022-10, Plastics - Determination of the melt mass-flow rate (MFR) and melt volume-flow rate (MVR) of thermoplastics - Part 1: Standard method (ISO 1133-1:2022): DIN-Normenausschuss Kunststoffe (FNK)

https://doi.org/10.1515/9783110785647-006

[31] ISO 11443:2021. Plastics - Determination of the fluidity of plastics using capillary and slit-die rheometers, 2021: ISO International Organization for Standardization
[32] DIN EN ISO 3219-1:2021-08. Rheology - Part 1, Vocabulary and symbols for rotational and oscillatory rheometry (ISO 3219-1:2021), 2021: DIN Deutsches Institut für Normung e. V.
[33] Schröder, T., Rheologie der Kunststoffe, 2018: Hanser Verlag
[34] Mezger, T., The Rheology Handbook, Vol. 5. 2020: Vincentz Network, Hannover, Germany
[35] Han, C.D., Rheology and Processing of Polymeric Materials, Vol. 1 Polymer Rheology. 2007: Oxford University Press, Inc.
[36] Viskosimetrie - Messung von Viskositäten und Fließkurven mit Rotationsviskosimetern. 2001/2008: DIN Deutsches Institut für Normung e. V.
[37] Rheometrie - Messung von Fließeigenschaften mit Rotationsrheometern - Teil 4: Oszillationsrheologie. 2016: DIN Deutsches Institut für Normung e. V.
[38] Dealy, J. M., Larson, R. G., Structure and Rheology of Molten Polymers. 2006: Hanser Verlag
[39] Ehrenstein, G.W., Rasterelektronmikroskopie REM-FEM-EDX. 2019: Hanser Verlag
[40] Schmidt P.F., Praxis der Rasterelektronenmikroskopie und Mikrobereichsanalyse, 1994: expert Verlag
[41] Günzler, H.; Gremlich, H.-U., IR-Spektroskopie - Eine Einführung. Vol. 4. 2033: WILEY-VCH
[42] Hesse, M., Meier, H.; Zeeh, M., Spektroskopische Methoden in der organischen Chemie. Vol. 7. 2005: Georg Thieme Verlag
[43] Frick, A.; Stern, C., Praktische Kunststoffprüfung. 2011: Hanser Verlag
[44] Tieke, B., Makromolekulare Chemie - Eine Einführung. Vol. 3. 2014: WILEY-VCH
[45] Plastics - the Facts 2022. 2022: Plastics Europe
[46] Kämpf, G., Characterization of Plastics by Physical Methods. 1986: Hanser Verlag
[47] Kern, M.; Trempler, J., Beobachtende und messende Mikroskopie in der Materialkunde. 2007: Brünne-Verlag
[48] Patzelt, W. J., Polarisationsmikroskopie. Vol. 2.1985: Leitz
[49] Schuth, M; Buerakov, W., Handbuch Optische Messtechnik. 2017: Hanser Verlag
[50] Frick, A., & Stern, C., C., Muralidharan V., Practical Testing and Evaluation of Plastics. 2019: WILEY-VCH
[51] Ehrenstein, G.W., Mikroskopie. 2020: Hanser Verlag
[52] DIN EN ISO 6721. Plastics - Determination of dynamic mechanical properties. 2019: DIN Deutsches Institut für Normung e. V.
[53] Scheirs, J., Compositional and Failure Analysis of Polymers. 2000: WILEY-VCH
[54] Christoph, R., Neumann, H.J., X-ray Tomography in Industrial Metrology: Precise, Economical and Universal. 2011: Verlag Moderne Industrie
[55] Data sheets. https://www.kern.de/de/technische-datenblaetter-kunststoffe. 12.06.2024: KERN GmbH
[56] Chemie Wirtschaftsförderung GmbH. CAMPUS-Datenbank. 2023: Altair Engineering GmbH
[57] https://plasticker.de/preise/pms.php?show=ok&make=ok&aog=A&kat=Regranulat, Plastic prices. 12.06.2024: New Media Publisher GmbH
[58] Trzeszczynski, J.; Huzar, E., Chromatographia, 2001: Springer
[59] Osswald, T. A., Menges, G., Materials Science of Polymers for Engineers, 2012: Hanser Verlag
[60] VDI 3822 Failure analysis - Fundamentals and performance of failure analysis. 2011: Beuth Verlag
[61] VDI 3822 -2.1.3 Failure analysis - Defects of thermoplastic products made of plastics caused by an unfavourable choice of material and by defects in the material. 2012: Beuth Verlag
[62] Kerkstra, R.; Brammer, S., Injection Molding Advanced Troubleshooting Guide. 2021: Hanser Verlag
[63] Pötsch, G., Micheli, W., Injection Molding. 2007: Hanser Verlag
[64] Chung, Chan I., Extrusion of Polymers. 2019: Hanser Verlag
[65] Rauwendaal, C., Understanding Extrusion. 2018: Hanser Verlag
[66] Grellmann, W.; Seidel, S., Polymer Testing. Vol 3. 2022: Hanser Verlag

[67]   Ehrenstein, G.W., Riedel, G., Trawiel, P., Thermal Analysis of Plastics. 2012: Hanser-Verlag

[68]   Hecht, T., Physikalische Grundlagen der IR-Spektroskopie. 2019: Springer

[69]   Baur, E.; Harsch, G.; Moneke, M., Werkstoff-Führer Kunststoffe. Vol. 11. 2019: Hanser Verlag

[70]   Wachs, I.; Banares, M., Advanced Catalyst Characterization. 2023: Springer

[71]   Gey, M., Instrumentelle Analytik und Bioanalytik, Vol. 4. 2021: Springer

[72]   Ortega, E., Hosseinian, H., Meza, I.; Lopez, M.; Vera, A., Hosseini, S., Material Characterization Techniques and Applications, 2022: Springer

[73]   Wampfler, B.; Affolter, S.; Ritter, A.; Schmid, M., Measurement Uncertainty in Analysis of Plastics. 2022: Hanser Verlag

[74]   Bauch, J.; Rosenkranz, R., Physikalische Werkstoffdiagnostik. 2017: Springer

[75]   Fedelich, N., Evolved Gas Analysis. 2019: Hanser-Verlag

[76]   Pastor, K., Emerging Food Authentication Methodologies Using GC/MS. 2023: Springer

[77]   Skoog, D.; Leary, J., Principles of Instrumental Analysis, 1992: Sounders College Pub.

[78]   aprentas, Laborpraxis Band 3: Trennungsmethoden, Vol. 6, 2017: Springer International Publishing

[79]   Leibnitz, E., Handbuch der Gaschromatographie, Vol. 3, 1984: Geest & Portig

[80]   Gross, J. H., Mass Spectrometry, 2017: Springer

Index

https://doi.org/10.1515/9783110785647-007